Science of Electricity

Volume 1
Introduction to Electrical Power

by Mark Fennell
© 2013

This book is part of the
Energy Technologies Explained Simply™ Series

Other Books in the Energy Technology Series

Paperback Books

Hydropower Technologies Explained Simply
Wind Power Technologies Explained Simply
Solar Power Technologies Explained Simply
Coal Power Technologies Explained Simply
Natural Gas and Other Hydrocarbon Technologies Explained Simply
Transmission of Electrical Power Explained Simply
Utility Operations and Grid Systems Explained Simply

e-Books

Practical Considerations of Solar Power
Advanced Solar Cell Technologies
Formation and Mining of Coal
Clean Coal Technologies
Mercury and Coal Power
Nuclear Power Meltdowns and Explosions
Health Hazards of Radioactive Decay
Radiation Measurements
Processes of Radioactive Decay and Storage of Nuclear Waste
Natural Gas Basics
Extracting and Refining Natural Gas (includes Fracking)
Transportation, Storage, and Use of Natural Gas
Introduction to the Transmission of Electrical Power
Power Lines
Underground Cables
Utility Operations and Quality Control
Power Grids

© Copyright notice
   This work is copyrighted to the author. No part of this book can be reproduced or sold for any reason without permission from the author.

About the Book

This book is the first in a series of books about energy technologies. In this book you will learn all the basic concepts of electrical power, including current, voltage, energy, turbines, generators, batteries, and electrical connections. The topics discussed in this book are fundamental concepts which apply to all books in the series.

In addition, you will learn some new fundamental concepts about electrical power. These are discoveries that the author has made from his extensive research and experimentation. These discoveries are unique to this book, they are published nowhere else.

After reading this book you will have a clear understanding of electrical power. You will understand how turbines and generators work. (This is essential knowledge for obtaining the greatest efficiency from power production). You will also understand your choices for batteries and electrical connections.

In this book you will learn how to read voltage and current sine wave graphs (an essential skill when you want to monitor the amount of power being transmitted). You will also learn the most common electrical terms and practical applications of each.

In addition, you will find several tables comparing units so that you may compare data from reports more easily.

Most important, you will understand all of these concepts easily. This book is designed for readers with little or no technical knowledge. Every concept is explained using simple language, a conversational style, and detailed diagrams.

About the Energy Technology Series

Purpose of this series

The books in the *Energy Technologies* series are designed to educate citizens, students, and legislators on all aspects of energy technologies. The first books in the series focus on electrical power.

The books discuss many energy technologies, including: generators, turbines, power plants, power lines, and grids. The technologies for each type of power source (hydro, wind, solar, coal, nuclear, and natural gas) are discussed in detail. The books also discuss efficiency, safety, reliability, and health concerns for each energy technology.

The ultimate goal of the series is to enable the people to make informed decisions on practical energy questions. The secondary goal is to serve as introductory guides for students embarking on careers with energy technologies.

Taken altogether, the books in the series answer any question you are likely to have, such as:
- How can we increase the efficiency of solar cells?
- How do I select the size my solar array?
- What do I need to know when installing a wind turbine?
- How effective are the clean coal technologies?
- How can we prevent grid failures?
- Do power lines cause cancer?
- and many other energy technology questions...

Science of Electricity in Perspective

The subject of electrical power is of great importance to our communities, but is rarely taught. Public debate is frequent and passionate, but with too little understanding of the actual science. At best, an informed citizen knows only a few pieces. At worst, as it is for a great number of citizens, electricity is magic, and myths are believed as scientific truth. It does not have to be that way. Any citizen, regardless of background, can know the technologies behind all aspects of electricity.

The books in this series solve that problem. These books educate the general public in all aspects of electrical power. Any person, regardless of background, can easily find the answer to his energy question in one of these books.

Specific Goals

There are numerous technologies described in these books. Yet for each technology I sought out the answers to the following questions:
1. How does the technology work?
2. What are the advantages and disadvantages?
3. What is the efficiency? How can the efficiency be improved?
4. What is the environmental impact? How can it be improved?
5. What are the safety hazards, and how can they be reduced?
6. What are the most important practical tips?
7. What facts comprise the most important data?

Technical Discussions Explained Simply

The books in the series must necessarily be technical to some degree. Electricity is a practical technology, and therefore we must understand the technical aspects if we want to make wise decisions. Yet the discussions in this book are always aimed at the citizen or policy maker.

The books in this series explain the principles of electricity as simply as possible, using ordinary English (no engineering jargon), and highlighting the most important points of each technology. Main concepts and facts are emphasized with the use of lists, tables, diagrams, and summaries.

I do not expect any reader to have a background in science, yet I offer enough facts and details so that the reader can have an accurate understanding of all related technologies. I provide enough technical details and enough data for the reader to make informed decisions.

Conclusion

For all the reasons above, I offer this series of books. Remember that there are no perfect solutions, there are only choices. I hope that this series of books will assist you in making those choices for your community.

Mark Fennell

Accuracy and Technical Depth

Objectivity

I have tried my best to be as objective as possible. Whereas many other authors of energy books have an agenda, I have no desire to promote one industry over another. I have no desire to promote one technical solution over another. In this endeavor, I have tried to be an objective scientist.

Accuracy of Data and Summaries

I never relied solely on the conclusions of other researchers. Instead, I performed many other tasks to ensure that all conclusions were accurate. I examined primary data whenever possible. I have read the fine print on how research was obtained.

I have also checked the accuracy of the conclusions written by other researchers, most commonly by finding at least three distinct sources for each fact. In addition, I performed my own calculations numerous times to prove (or disprove) conclusions and final values found in other reports. It is only after such rigorous investigations that I created data tables and wrote summaries for these books.

Limited Mathematics

The books must also use math from time to time. For example, efficiency is a statement of a specific amount, and therefore the discussion of efficiency requires the use of equations. Other issues such as power loss, health hazards, environmental concerns, and quality control are also statements of amounts and also require calculations. Therefore some equations are necessary to know, even for the non-scientist.

I also provide examples of calculations so that readers can become more comfortable with using the equation themselves.

However, I want to emphasize that I focus on concepts not on the mathematics. I provide equations only when it is necessary for the citizen or student to be familiar with these equations.

M.F.

Table of Contents

1. Basic Principles of Electrical Power	11
2. Operation of Turbines and Generators	25
3. Voltage, Current, and Power on the Atomic Scale	43
4. Alternating Power and Frequency	49
5. Voltage & Current Sine Wave Graphs	53
6. Batteries	59
7. Resistance and Temperature	77
8. Electrical Connections: Series and Parallel	85
9. Other Primary Electrical Terms	91
10. Impedance and Reactance Details	99
Conclusion	105
Appendices	107
Bibliography	112
Index	113

Table of Contents: Detailed

1.1 Basic Principles of Electrical Power   11
1. Main Types of Electrical Plants
2. Energy
3. The Joule
4. The Kilowatt-Hour
5. Power
6. Electrical Power, Current, and Voltage
7. Overview of Turbines and Generators

1.2 Operation of Turbines and Generators   25
1. Turbines Pushed by Molecules
2. Creating a Flow
3. Non-Conversion Turbines: Water and Wind
4. Conversion Turbines: Steam Turbines and Gas Turbines
5. Dual Turbines
6. Operation of Generators: Overview
7. Magnets and Magnetic Fields
8. Creating Electrical Current
9. Simplified Generator
10. Alternating Current
11. Broad Magnetic Field
12. Graphing Alternating Current
13. Three Phase Electrical Supply

1.3 Voltage, Current, and Power on the Atomic Scale   43
1. Electrical Current
2. Voltage
3. Electrical Power
4. Technical Discussions on Current
5. Technical Discussions on Voltage

1.4 Alternating Power and Frequency   49
1. Alternating Power Basics
2. Frequency Basics
3. Alternating Voltage and Alternating Current
4. Forward Power vs. Alternating Power

1.5 Voltage & Current Sine Wave Graphs   53
1. Voltage and Current Each Graphed by Sine Waves
2. Understanding Voltage Sine Wave Graphs
3. Understanding Current Sine Wave Graphs

1.6 Batteries   59
1. Battery Power versus Turbine-Generator Power
2. Overview of How Batteries Work
3. Anode and Cathode Overview
4. Electrolyte Solution to Create Free Electrons
5. Pull Force of Nuclei and Voltage of Batteries
6. Corrosion as Related to Anode and Cathode
7. Life Cycle of Batteries
8. Rechargeable vs. Not Rechargeable
9. Specific Differences between Battery and Generator
10. Battery Use for Solar Power
11. Size of Batteries
12. Depth of Discharge Overview
13. Depth of Discharge in Detail

1.7 Resistance and Temperature   77
1. Resistance and Temperature Basics
2. Resistance and Temperature on the Atomic Scale
3. Resistance and Heat on the Atomic Scale
4. Factors Affecting Resistance and Temperature
5. Power vs. Power Loss
6. Resistance, Heat, Power Loss, and Temperature

1.8 <u>Electrical Connections: Series and Parallel</u>　　　85
1. Appliances Connected in Series and Parallel
2. Power Sources Connected in Series and Parallel

1.9 <u>Other Primary Electrical Terms</u>　　　91
1. Conductor, Resistor, Resistance, Insulator
2. Inductor, Inductance, Capacitor, Capacitance
3. Rotor, Stator, Electric Motor
4. Conversion Equipment

1.10 <u>Impedance and Reactance Details</u>　　　99
1. Impedance: Basic Concepts
2. Resistance, Reactance, and Impedance Specifics
3. Symbols: L, C, R, X, and Z
4. Calculations of Reactance and Impedance

Conclusion　　　105
Appendices　　　107
Bibliography　　　112
Index　　　113

1.1
Basic Principles of Electrical Power

Introduction

Electrical power can be viewed in several ways. Electrical power can be considered to be a combination of the energy of electrons and the flow of those electrons. Electrical power can also be viewed as a rate, such as the rate of electrical energy produced per time. Electrical power is most often said to be a combination of voltage and current.

Electrical power is created by a series of turbines and generators. The generator is the actual device which creates the electricity. However, turbines are necessary to create the rotation required for the generator.

There are many types of electrical power plants. However, most electrical power plants are similar in the following set of steps:
1. The stored energy must be converted into a flow.
    (This is usually a flow of water, steam, or air.)
2. The flow of the molecules operates the turbine.
3. The turbine operates the generator.
4. It is the generator that actually creates the electricity.

The differences of each type of power plant depend primarily on the method of converting the stored energy (such as in a piece of coal) into a flow (such as the flow of steam). At that point, all power plants are similar: the flow of the molecules, such as steam, operates the turbines. These turbines operate the generators, which then produce the electricity.

List of topics in this chapter
1. Main Types of Electrical Plants
2. Energy
3. The Joule
4. The Kilowatt-Hour
5. Power
6. Electrical Power, Current, and Voltage
7. Overview of Turbines and Generators

Main Types of Electrical Plants

The main types of electrical power plants are explained briefly below. The details will be discussed in subsequent books.

1. Hydroelectric Power: Water is stored in a reservoir above a dam. The water flows down a channel and pushes the blade of the turbine. The turbine rotates, which then operates the electrical generator.

2. Wind: Wind is already in a state of flow, so there is nothing to convert. Thus, the wind hits the blades of the turbine, and the rotating turbine operates the electrical generator.

3. Solar: There are several ways to use power from the sun. The most common is to create electricity by using a photovoltaic cell. In a photovoltaic cell, the sun excites electrons in the material of the cell. The solar cell is designed to force the excited electrons into a direct current. The current is then converted from direct current (DC) into alternating current (AC). The alternating current is then sent through traditional wiring.

4. Hydrocarbons: Hydrocarbon fuels include coal, diesel, and natural gas. Coal is the most common fossil fuel for large power plants. Natural gas is ideal for power plants which are used only at times of peak use. Diesel is used to operate smaller generators, particularly for emergency use and remote locations.

For any type of hydrocarbon fuel, the fuel is burned which then creates heat. This heat is applied to a water supply, thereby turning the water into steam. The flow of this steam then pushes the blades of the turbine. The rotating turbine then operates the electrical generator.

5. Nuclear Power: Nuclear power is based on the energy stored in the nucleus of an atom. Splitting the nucleus of an atom apart will unleash great energy. This energy is used to heat water, thereby turning the water into steam. This steam flows to the turbines and pushes the blades of the turbine. The rotating turbine operates the electrical generator.

6. Biomass and Trash: Biomass is essentially wood and dead plants. Biomass can be burned which creates heat. The heat turns water into steam, the steam operates the turbine, which operates the generator, which creates electricity. Similarly, trash can be burned, creating heat. The heat turns water into steam, the steam operates the turbine, the turbine operates the generator, and the generator creates electricity.

Energy: Concepts and Units

Introduction

Energy is an intuitive concept. However, in order to measure energy and compare values we must understand energy in exact terms.

Energy can be best defined as "the ability to move an object or to change an object's physical properties". Energy is also defined as "the ability to do work". Common energy units include the Joule and the BTU.

Common Energy Units and Abbreviations

1. Calorie, cal
2. Joule, J
3. British Thermal Units, BTU
4. electron volt, eV
5. Megaelectron Volt, Mev
6. kilowatt-hour, kw-hr (or kWh)
7. Watt-hour, Wh

Comparison Guide for Energy Sizes

In order to give you a feeling for the relative sizes of the various energy units, I have created the following list. The units are arranged in the order of size, and compared to the common unit of the Joule. Note that you can use these numbers for rough estimates but not for exact calculations. For exact conversions, see the Appendix.

Unit	Equivalent in Joules (approximate)
eV	a millionth of a trillionth of a Joule
MeV	a trillionth of a Joule
Joule	1 Joule
calorie	4 Joules
BTU	1,000 Joules
kw-hr	3.6 million Joules (3,600,000 Joules)

Also useful as quick energy comparisons:
- A BTU is about 1,000 times greater than a Joule
- A kw-hr is about 3,600 times greater than a BTU

Definitions of the Common Energy Units

1. <u>Calorie</u>: the energy required to raise the temperature of one gram of water one degree Celsius.

2. <u>Joule</u>: the energy required to move an object with the force of one Newton through a distance of one meter: 1 Joule = 1 Newton x 1 meter. The Joule is the most universal energy unit.

Note that a Joule can also be related to kilogram, meters, and seconds. This is because the Newton is actually related to kilogram, meters, and seconds: 1 Newton = 1 kg x m/s². Therefore a Joule can be related to kilogram, meters, and seconds as follows:

1 Joule = 1 Newton x 1 meter = [1 kg x m/s²] x 1 m = 1 kg x m²/s²

3. <u>British Thermal Units</u>: the energy required to raise the temperature of one pound of water one degree Fahrenheit.

4. <u>electron volt</u>: the energy gained by an electron when it increases one volt in potential.

5. <u>Watt-hour</u>: the energy equivalent of the power of one watt operating for one hour. Note that this definition is created in reverse, starting with the power unit of Watt (see later in this chapter).

Related is the kilowatt-hour, which is the energy equivalent to the power of one kilowatt (1,000 Watts) operating for one hour. The kilowatt-hour is the unit of energy commonly used by power companies in computing your electric bill.

The Joule

Introduction

The unit of the Joule is the most universal of all the energy units. Therefore, we should have a more complete understanding of the Joule. Repeating from above: 1 Joule = 1 Newton x 1 meter = 1 kg x m²/s².

The data table below shows the size of the Joule in relation to common electrical devices. Regarding the data below note that appliances vary, such as the variety of televisions available for sale. Therefore the numbers used in the table below are taken as the average of the most common devices. For example, the value of Joules for "television" in the data below is taken as the average energy requirement for the common televisions available today.

The Number of Joules to Run Devices for One Hour

Appliance	Joules Required for 1 hour of use
1. Light bulb	162,000 Joules
2. Television	1,080,000 Joules
3. Refrigerator	2,160,000 Joules
4. Computer	3,600,000 Joules
5. Dishwasher	10,800,000 Joules
6. Air Conditioner	18,000,000 Joules

The Kilowatt-Hour

The kilowatt-hour deserves special attention. The kilowatt-hour is a unit of energy (not power). It is derived in reverse, from the definition of kilowatt. (See the section on Power below.) The utility company uses the kilowatt-hour as a simple way of figuring how much energy you used, then bill you accordingly.

Also note that because the Watt is related to the Joule, then:

1 kilowatt-hour = 3.6 million Joules.

For more information on the Watt, see the section on power later in this chapter.

Power: Concepts and Units

Introduction

Power is energy with a time component. Power is a rate. Therefore power is the amount of energy produced in a period of time. Power is also the amount of energy *used* over a certain period of time.

The most common units of power are watts and kilowatts. A watt is the number of Joules of energy created or used per second. This is somewhat of a small number, so kilowatts are preferred. A kilowatt is the number of kilojoules of energy created or used per second.

Common Power Units and the Abbreviations

1. Watt, W
2. kilowatt, kW
3. calories per second, cal/sec
4. British Thermal Units per hour, BTU/hr
5. Horsepower, H.P.
6. Volt-amperes
7. electron Volts per sec, eV/sec

Comparison Guide for Power Units

In order to get a feeling of power unit sizes, I have created a list below. The units are arranged from the unit of smallest size to the unit of largest size, and then each is compared to the common unit of watts.

Note that some power units work with seconds, while other power units work with hours. Also note that these values are approximate in order to simplify comparisons. For exact values and conversions see the Appendix.

Unit	Equivalent in Watts (Joules/sec)
BTU/hr	.3 Watts
Watt (J/sec)	1 Watt
cal/sec	4 Watts
Horsepower	745 Watts
kilowatts (kJ/sec)	1,000 Watts

Also useful as quick comparison of power units:
- Each BTU/hr is about 1/3 of a Watt.
- Or, each Watt is about 3 times greater than a BTU/hr
- One Horsepower is about 3/4 of a kilowatt.

Definitions of the Common Power Units
1. Watt: the amount of power expressed in Joules of energy per second.
          1 Watt = 1 Joule /second.

Technically, the Watt is the amount of power required to move an object with the force of one Newton through a distance of one meter, in one second. Therefore, the watt can also be expressed in terms of kilograms, meters, and seconds. 1 Watt = 1 kg x m²/s³ (Also refer to the definition of Joule above). The watt is the most universal of power units.

2. kilowatt: the amount of power expressed in kilojoules of energy per second. One kilowatt = one kilojoule of energy per second.

3. cal/sec: the amount of power required to raise the temperature of one gram of water one degree Celsius, in one second.

4. BTU/hr: the amount of power required to raise the temperature of one pound of water one degree Fahrenheit, in one hour.

5. Horsepower: the amount of power required to move an object weighing one pound a distance of one foot, in one second. (This is an older unit and is only given here to assist when referencing historical records.)

6. Volt-amperes: another power term equivalent to Watts. However, this unit is used when making the distinction of a different type of power. We will discuss this later in the chapter on Power Factor in the Unit on Efficiency.

7. electron Volts per second: the amount of power required to raise the potential of one electron by one volt, in one second. Note that these units are small, and therefore are not often used in larger calculations.

Electrical Power, Voltage, and Current

Introduction

Electrical power is a combination of voltage and current. This relationship is the bottom line, most fundamental relationship for electrical power. All of the things that we do regarding electricity are related to that relationship.

> In concept: Power = Voltage x Current
> In units this means: Watts = Volts x Amps
> In symbols this is: P = V x I

Current is the flow of electrons through a wire, and is measured in Amps. The value of one Amp is approximately $6.5 \times 10^{+18}$ electrons traveling through a wire per second. Different wires will carry different amounts of current. Larger wires can carry more current (more Amps). When you read the warranty or installation instructions for your device, that document will usually tell you what amperage wires to use.

Voltage is the overall energy of many electrons at one location. Voltage is measured in Volts: 1 Volt is equal to 1 Joule of energy per $6.5 \times 10^{+18}$ electrons. Your home is usually wired with a particular voltage. In many homes that voltage is 120 Volts. Many recent residential developments use 130 Volts. This voltage is determined by the transformer closest to your home.

As stated earlier, electrical power is a combination of voltage and current. Electrical power is most commonly measured in watts or kilowatts. For any device, the manufacturer will tell you how much power, in Watts, is required to run that device.

Examples

1. Your home is wired with 120 Volts. Your device uses 10 Amps of current. Electrical Power = 120 Volts x 10 Amps = 1,200 Watts.

2. A typical refrigerator wire requires 7.0 Amps of current. Most homes are given 120 Volts by the utility company. Therefore, the power required to operate the refrigerator = 120 Volts x 7.0 Amps = 840 Watts.

3. A light bulb manufacturer provides the following information on its packages, as comparative information:

Volts	Watts
130 Volts	75 Watts
120 Volts	69 Watts

The information tells us that we can still use the bulb if we have only a 120 Volt line in the house. However, the bulb will be dimmer in a 120 Volt house than in a 130 Volt house. The bulb is the same in each home, approximately .57 Amps. However, since the voltage is different in the differently wired homes, then the power output of the same bulb will be different. A more complete table of information would be:

Volts	x	Amps	=	Watts
130 Volts		.577 Amps		75 Watts
120 Volts		.577 Amps		69 Watts

Thus, if my home is wired to give me 120 Volts then I will get a lower amount of power from my bulb, that of 69 Watts. However, if my home is wired to give 130 Volts then I will get a higher amount of power from my bulb, that of the full 75 Watts.

Turbines and Generators: Overview

All sources of energy (such as hydroelectric, coal, or wind) must be transformed into electrical power. This is where turbines and generators come into use.

Without turbines and generators, we would not have any electrical power. Therefore, if we wish to understand energy technologies, and if we wish to make wise choices regarding those technologies, then we must first understand turbines and generators.

    a. Turbines: a turbine is essentially an efficient water wheel. We need the turbines in order to operate the generators.

    b. Generators: the generator is the actual device that creates the electricity, regardless of the original energy source. The generator itself is a system of magnets and electrical wires. Rotating these magnets creates electricity.

Turbines

A turbine is an efficient water wheel. In fact, a good definition for a turbine is a machine that converts straight line movement (such as the flow of water) into rotational movement (such as the rotating pole attached to the water wheel). Note that we need to create rotational movement because the generators need rotational movement in order to create electricity. Hence we need the turbines.

Water wheels have been used for centuries, such as in lumber mills and flour mills. In these water wheels, the flow of water from a river pushes the blades on the wheel, causing the wheel to rotate. The wheel is then connected to some other device, which is used for practical purposes such as for milling. Thus, the kinetic energy of flowing water is transformed into mechanical energy, which is then used to operate a mechanical device.

The turbine uses the same principle. Turbines have blades, just as a water wheel does. The flow of water hits the blades of the turbine, and pushes the turbine in circles. Also, as with the water wheel, the turbine is connected to a mechanical device. The turbine uses that mechanical energy for a practical purpose. However, if we want electricity then we do not want just mechanical energy, we want electrical energy, and so we have to proceed one step further. That next step is the electrical generator.

Generators, Magnets, and Electrical Current

<u>Introduction</u>

Generators are the actual devices which create the electricity. A generator needs three things to create current: 1) a magnet, 2) a wire, and 3) the relative movement between the magnet and the wire. When a magnet moves past a wire, or when a wire moves past a magnet, then electricity is generated.

The specific actions which create the electric current are subtle. The specific mechanisms involve magnetic fields, electron spins, and flow of electrons.

<u>Magnets, Magnetic Field, and Creation of Electrical Current</u>

Note that we will discuss the process of using magnets to create electrical power in much greater detail in the chapter on Turbines and Generators. At this point we will have a brief overview.

What is a magnet? A magnet is any material where the majority of electron spins are aligned. This creates a type of energy known as the magnetic field.

The magnetic field energy not only flows in the magnet, but also surrounds the magnet by a distance of several inches or feet. Therefore the magnetic field can reach any wire sitting nearby. Then, because the magnetic field extends beyond the borders of the magnetic material, this field will impart energy to the electrons on the nearby wire.

A magnetic field will impart energy to electrons in any metal. The amount of energy applied to the electrons will depend primarily on three factors: the strength of the magnetic field, how close the magnetic field is to the electrons on the wire, and the atomic structure of the metal in the wire.

However, if the magnet and metal are stationary then the amount of current is minimal. The energies may be absorbed in other ways besides forward motion. (These other ways include electron vibration, electron spin, and other motions). Consequently, the forward motion of electrons may not be enough for practical use.

Therefore, we move the magnet (and move the magnetic field energy) which will then induce an electrical current in a nearby wire. This is the basic operation of how the generator creates electrical current.

Magnetic Field and Alternating Current

Note that the direction of the magnetic field will push the electrons on our wire to flow in a particular direction. Therefore, if we want to create alternating current then we must induce the current in two directions. This is done easily enough by rotating the magnet – and hence rotating the magnetic field. Rotating the magnet pushes the electrons in one direction, and then pulls the electrons in the opposite direction. Using this basic system, we can create alternating current.

Turbine and Generator Together

Now we can understand the turbine and generator together. The turbine rotates. This turbine is connected to the magnet in the generator. Thus, as the turbine rotates, the magnet also rotates. Then as the magnet rotates past the wire, electricity is created.

Also, because the direction of the magnetic field dictates the direction of the electrical current, as the magnet rotates the direction of the electrical current will change accordingly. The continuous rotation of the magnet will therefore create a continuous alternating current.

Chapter Summary

1. Most electrical power plants are similar in the following set of steps:
    a. The energy must be converted into a flow,
        usually a flow of water, steam, or air.
    b. The flow of the molecules operates the turbine.
    c. The turbine operates the generator.
    d. It is the generator that actually creates the electricity.

2. The difference in each power plant type is in the method of converting the stored energy (such as in a piece of coal) into a flow (such as the flow of steam).

3. Energy is defined as "the ability to move an object", "the ability to change an object's physical properties", or "the ability to do work".

4. Common units for energy include: calorie, Joule, BTU, eV and kilowatt-hour.

5. The unit of the Joule is the most universal. A light bulb requires approximately 50 Joules to operate for 1 second, and approximately 170,000 Joules to operate for an hour.

6. The sizes of units of energy can be compared in approximate amounts as follows:

Unit	Equivalent in Joules (approximate)
ev	a millionth of a trillionth of a Joule
Mev	a trillionth of a Joule
Joule	1 Joule
calorie	4 Joules
BTU	1,000 Joules
kw-hr	3.6 million Joules (3,600,000 Joules)

7. Power is the amount of energy produced or used over a period of time. Power is a rate.

8. The most common units for power in electricity are watt and kilowatt. 1 watt = 1 joule of energy per second. 1 kilowatt = 1 kilojoule (1,000 Joules) of energy per second. Other units that are used include: cal/sec, BTU/hr, eV/sec, and horsepower.

9. The sizes of units of power can be compared in approximate amounts as follows:

Unit	Equivalent in Watts (Joules/sec)
BTU/hr	.3 Watts
Watt (J/sec)	1 Watt
cal/sec	4 Watts
Horsepower	745 Watts
kilowatts (kJ/sec)	1,000 Watts

10. Electrical Power = Voltage x Current.
    In concept: Power = Voltage x Current
    In units, this means: Watts = Volts x Amps
    In symbols, this is: $P = V \times I$

11. Current is the flow of electrons. The rate of current is measured in Amps. 1 Amp = $6.5 \times 10^{+18}$ electrons traveling per second.

12. Voltage is the overall energy of many electrons at one location. Voltage is measured in Volts. 1 Volt is equal to 1 Joule of energy per $6.5 \times 10^{+18}$ electrons.

13. The Voltage is fixed in your home, usually 120 or 130 Volts. The voltage is determined by the transformer closest your home.

14. The amount of current (measured in Amps) is determined by the wire size of each particular appliance. Larger wires carry more current.

15. Electrical power (measured in Watts) is a combination of the fixed voltage in the walls and the wiring in the particular appliance.

16. Turbines transform a straight flow (of water, wind, steam, or gas) into a rotational movement. This rotational movement is needed to operate the generators.

17. Generators are the devices that actually create the electricity. Electricity is created by the relative movement between magnets and wires within the generator.

1.2
Operation of Turbines and Generators

Introduction

As stated earlier, turbines convert the straight flow of something (such as steam or wind) into rotational movement. The rotation of the turbine is necessary to rotate the generator.

Generators are the actual devices which create the electricity. The generator requires three things to create current: 1) a magnet, 2) a wire, and 3) the relative movement between the magnet and the wire. When a magnet moves past a wire, or when a wire moves past a magnet, electricity is generated.

The specific turbine and generator process is as follows: The energy that we need arrives at the turbine as flowing molecules. These molecules push on the turbine blades, causing the turbine to rotate. The turbine is attached to a magnet, so that as the turbine rotates the magnet also rotates. The rotational movement of the magnet creates the electricity.

List of Topics for this chapter
1. Turbines Pushed by Molecules
2. Creating a Flow
3. Non-Conversion Turbines: Water and Wind
4. Conversion Turbines: Steam Turbines and Gas Turbines
5. Dual Turbines
6. Operation of Generators: Overview
7. Magnets and Magnetic Fields
8. Creating Electrical Current
9. Simplified Generator
10. Alternating Current
11. Broad Magnetic Field
12. Graphing Alternating Current
13. Three Phase Electrical Supply

Turbines Pushed by Molecules

Note that technically all turbines are pushed by molecules. The most common molecules used to push turbine blades are water, wind, steam, and gas.

Water turbines: In water turbines, water molecules in the liquid phase push the turbine blade.

Wind Turbines: In wind turbines, air molecules (primarily nitrogen and oxygen in the gas phase) push the blades.

Steam Turbines: In steam turbines, water molecules in the gas phase push the turbine blades.

Gas Turbines: In gas turbines, there are carbon dioxide molecules, water molecules and some methane molecules, all of which are in the gas phase, which push the turbine blades.

Creating a Flow

We must first create a flow of energy from stored energy. Some sources of energy are already in a state of flow, while other sources must be converted from stored energy to flowing molecules.

Water is both a source of energy as well as the flow of that energy. (Stored water is a source of energy; falling water is the flow of that energy). Similarly, wind is both a source of energy and the flow. (Air molecules are the source of energy; wind is a flow of that energy). Therefore, for water power and wind power we do not need to convert anything. The water, or the wind, is applied as a flow directly to the turbine blades.

However, with steam and gas turbines we have to create the steam or create the gas. Steam and gas are not the source of energy; the actual sources of energy include coal, nuclear pellets, and biomass. In these cases, the energy sources must be converted into the steam or converted into the gas, so that we can then apply the flow of molecules to the turbine blades.

Non-Conversion Turbines: Water and Wind

Water and wind do not require conversion before sending to the turbine. Therefore, the turbines are relatively simple. However, there are many ways to improve the efficiency of these turbines.

We want turbines to rotate as fast as we can, because the speed of rotation is directly related to power. In some situations, such as wind power and micro-hydro power, the turbine alone does not rotate fast enough to get the power that we need. In these cases, we need to boost the speed of the rotating axel before it reaches the generator. We boost the speed through a series of gears. Note that belts and pulleys, instead of gears, can be used for the same purpose.

You will read data on gears in a form such as 4:1. This means that the output rotational speed is 4 times faster than the rotational speed to begin with.

Basic Water Turbine:
The Classic Water Wheel

Basic Wind Turbine:
The Classic Windmill

Conversion Turbines: Steam & Gas Turbines

Introduction

We have two distinct types of conversion turbines, each of which are based on converting energy stored in a material into the flow of molecules. These two turbines are the steam turbine and the gas turbine. We will look at the major principles of these two turbines in the following sections.

Steam Turbine (Figure 2.1)

Steam turbines are used where the source of energy is any of the following: coal, nuclear, biomass, oil, or natural gas. Steam turbines are also used where solar power is created by using mirrors or lenses.

The steam turbine is based on boiling water. First we release energy from a material, such as by burning coal. After we release the energy, we apply that energy to water. This water is usually in a pipe which runs through the furnace. The energy created in the furnace heats the water, which makes the water boil, and converts the water into steam. The steam is the actual substance which pushes the turbine blades.

In a steam turbine the water can be reused over and over again. This happens as follows: After pushing the turbine blades, the steam flows into a cooling tower, condenser, or pond. Most often, the steam is simply kept in a storage area and allowed to cool back down into water.

Steam Turbine   Figure 2.1

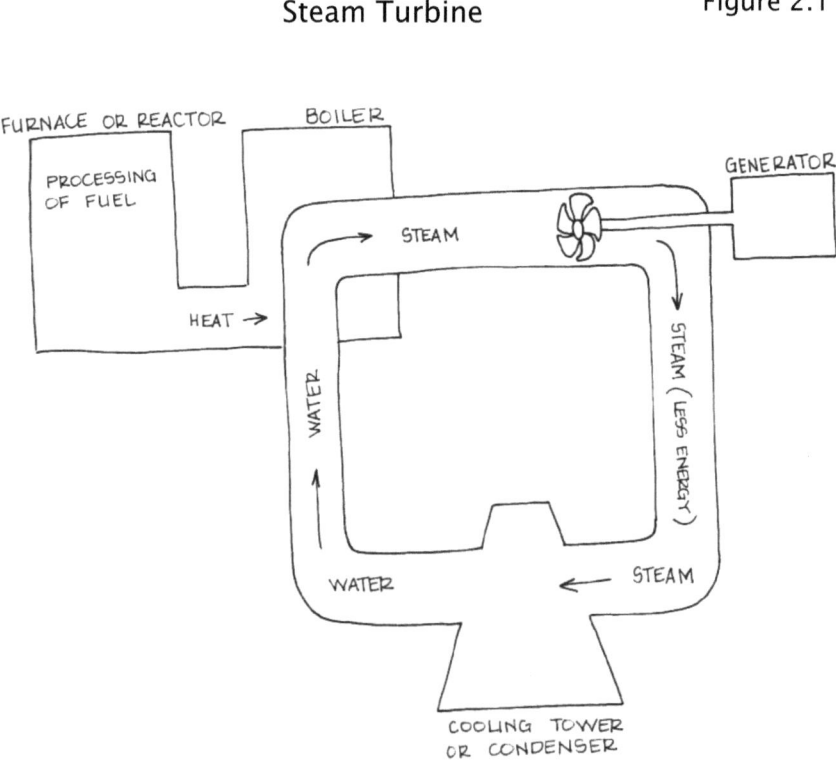

Alternate Cooling Method

An alternate method for cooling is to send a second source of water past the steam via a second set of pipes. The cooling water absorbs the heat, while the original turbine water is cooled down and can be reused.

Note that this coolant water (in the alternate cooling method) is often taken from local rivers and lakes. This water is not really "used" because it is sent right back into the river. However, hot water can kill the fish, so any coolant used in power plants must itself be allowed to cool before returning to the river. Many power companies have moved toward having their own self-contained water supplies, which allows the company to operate without ever touching the local ecosystem.

Gas Turbines (Fig 2.2)

Gas turbines are used most often in coal power plants. In a gas turbine, molecules in the gas phase are the substances which actually push the turbine blades. The gas molecules are actually by-products of the process of burning.

The fuel used for burning is usually a hydrocarbon, such as wood, coal, or gas. When any of these fuels are burned we get carbon dioxide, methane, and water molecules, all of which are in the gas phase. These molecules are sent directly to the turbine and push on the turbine blades.

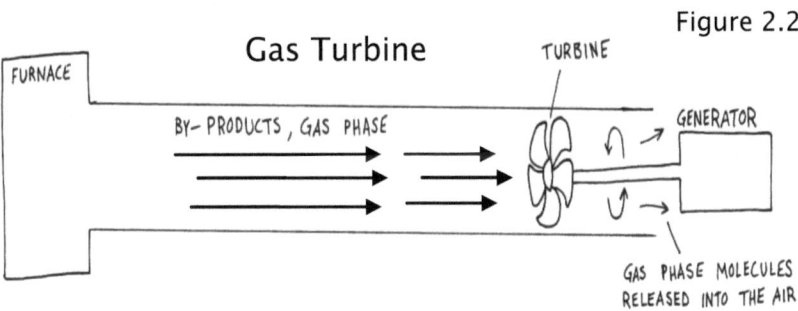

Figure 2.2

Note that some people will refer to this as a "combustion turbine." What these authors mean by combustion turbine is simply this: burning takes place (combustion), which produces molecules in a gas phase. The gas phase molecules then push the turbine blades.

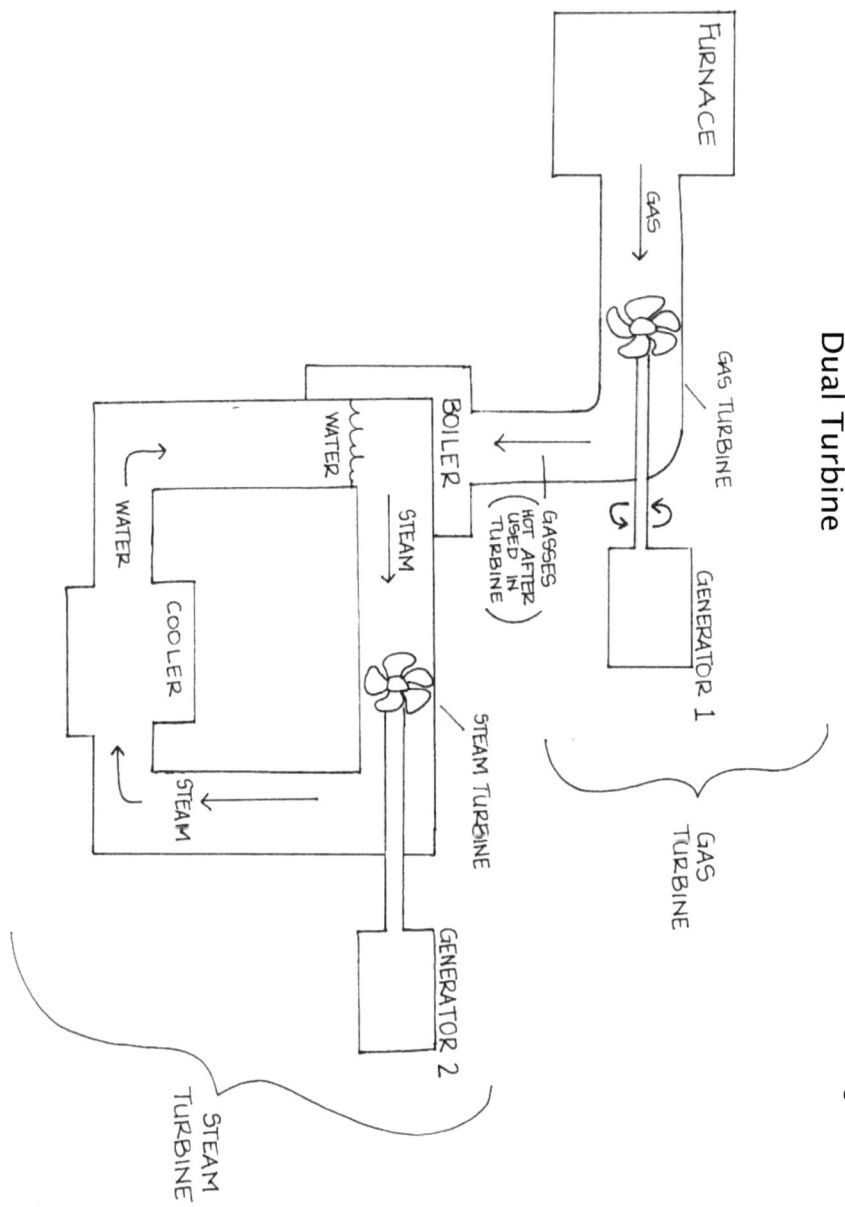

Dual Turbine

Figure 2.3

Dual Turbines (Fig 2.3)

Introduction

A dual turbine is any turbine system where two individual turbines are connected together. A dual turbine system is able to use most of the released energy from any given material.

There are several versions of dual turbines, depending on what turbine systems are connected. The most common Dual Turbine is the Combined Cycle Gas Turbine. Note that the Combined Cycle System is also called the Integrated Gasification Combined-Cycle (IGCC) Turbine.

Combined Cycle Gas Turbine (Figure 2.3)

A combined cycle gas turbine uses both the gas turbine and the steam turbine in the same system. In coal power, this is often called the IGCC turbine: the Integrated Gasification Combined-Cycle turbine.

The first turbine is a gas turbine, where gas molecules are created by the combustion, and these gas molecules push the turbine blades. Then, there is often enough energy leftover from the first turbine to be used elsewhere.

The second turbine is a traditional steam turbine, where the leftover energy of the gas molecules from the first turbine can actually heat water into steam. The steam then pushes the turbine blades of this second turbine. Thus, in the combined cycle turbine two turbines are used, and we get more power from the same amount of energy source.

Overview of Generators

Remember that we need three things to create current: a magnet, a wire, and the relative movement between the magnet and the wire. When a magnet moves past a wire, then electricity is generated.

The magnet is any material where the majority of electron spins are aligned. This alignment of electron spins creates a type of energy known as the magnetic field. The magnetic field is an energy which surrounds the magnet by a distance of several inches. Therefore the magnetic field can reach any wire sitting nearby. The energy of the magnetic field will therefore impart energy to our wire.

In addition, the direction of the magnetic field will push the electrons on our wire, creating a flow in a particular direction. Therefore, if we want to create alternating current then we must induce the current in two directions. This is done easily enough by rotating the magnet, and hence rotating the magnetic field. Rotating the magnetic field will push electrons in one direction, then pull electrons in the opposite direction. This creates an alternating electrical current.

Therefore, we use a rotating magnet, and its changing magnetic field energy, to induce a current in a nearby wire, in two directions. This is the basic operation of the generator.

Magnets and Magnetic Fields

Magnet

What is a magnet? In any atom, the electrons (and subatomic particles in the nucleus) have certain spins. These electron spins create a certain type of energy: magnetic energy. In most types of material these spins are random, and therefore cancel each other out. However, when many spins in a region of material are aligned in the same direction, then their energy combines. In total, the energy of these aligned electron spins will create a strong flow of energy in one direction. Therefore, any material we call a magnet has aligned electron spins, which will create a net flow of energy in one overall direction.

Magnetic Fields and Energy

The magnetic field is a type of energy. As discussed earlier, the magnetic field is produced by aligned spins of electrons and other sub-atomic particles. Most important, this energy field actually extends beyond the borders of the material.

The strength of a magnetic field (the strength of that energy) depends on the number of sub-atomic particles (primarily electrons) which have their spins aligned. Depending on the size and strength of the magnet, this field can extend inches, feet, or miles from the original object. It is this magnetic field which induces current in our wires.

Also note the following correlation: magnetic fields are created by electron spins, and we use the magnetic fields to apply energy to electrons in our wires. Therefore, we are in fact applying the energy from one group of electrons to another group of electrons, through the mechanism of a magnetic field.

Poles of Magnets

Every magnet has two poles. One pole is the direction all the energy is headed towards. This is usually termed the "north pole". The other pole is the direction in which all the energy is moving from, usually termed the "south pole".

Note that the north and south poles exist because of the directions of electron spins. The electrons will spin primarily in two directions, and therefore the resulting magnetic field energy will flow primarily in two directions.

When you place two magnets together, they will join if the poles are facing the same direction. This is because the energy of both magnets are flowing the same direction. I call this the "Bing Crosby" alignment, after the song "Going My Way". In the song, Crosby says "If you are going my way, then you can go my way, too". That is the nature of two magnets with their poles aligned – all the energy going "my way".

More specifically, when two magnets are brought together, facing the same direction, then the energies combine. The energy of the front magnet pulls on the energy of the back magnet. At the same time the energy of the back magnet is itself flowing the same direction as the front magnet. This is what makes the powerful pull you feel between two magnets, and what makes them snap together with such a strong force.

Similarly, if you place a strong magnet near a piece of metal (any piece of metal, even if not technically a magnet) then the energy of the magnetic field will interact with the electrons in that metal. The magnetic field is strong enough to physically push the electrons in one direction. This is how we use the magnetic field to create electrical current in a wire.

Creating Electrical Current

Therefore, this is what happens when we create electric current in a generator: We place a strong magnet next to our wire. The magnet has aligned electron spins, which create an automatic energy flow in one direction.

In addition, this energy flow (magnetic field) extends several inches beyond the regions of the magnet. This energy field then flows into the neighboring wire. Once there, the magnetic field imparts its energy to the electrons. Now the electrons have their energy. These energized electrons can then move in such a way as to create an electrical current.

As we change the orientation of the magnet, the direction of the magnetic field will change, and therefore the direction of energy flow imparted to electrons in the wire will change. In order to change directions of our magnet, we usually rotate it in a continuous circle. This is the basic action of the generator.

Simplified Generator

Creating electrical current requires relative movement between a magnet and a wire. Therefore, generators can be made two ways: a) the magnet moves, while the wires are stationary, or b) the wires move, while the magnet is stationary. For my examples, I will stay with one type of generator: the magnet rotates and the wires are stationary. However, both types of generators are used.

In order to understand these principles, we will use a simplified generator (Fig. 2.4). This simplified generator has a rotating magnet and one wire. The magnet is a straight magnet, with a North and a South pole. The middle of this magnet is attached directly to the turbine. Thus, as the turbine rotates, the magnet also rotates. The magnet then sweeps in a circle.

We then place one wire next to the magnet. When the rotating magnet sweeps past the wire, that is when electricity is created. As the magnet rotates, it sweeps past the wire twice every cycle (once when the North end passes by, once when the South end passes by). Each time that either the North or South end of the magnet passes the wire, electricity is produced. Creating electricity in this manner is called "induction."

Alternating Current

In a generator, the magnet rotates and passes the wire. Every time the North end passes the wire, current is produced. Every time the South end passes the wire, current is also produced, but in the *opposite* direction. Thus, with every cycle of the magnet, two currents are produced, in two different directions. This is alternating current.

Figure 2.4: Simplified Generator

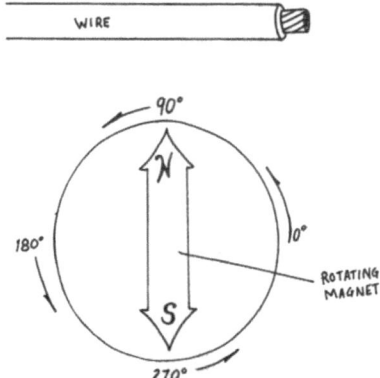

Broad Magnetic Field

Note that this description is a bit of a simplification. The reality involves the nature of the magnetic field. Although the magnet itself has a point, the magnetic field is broad, not a point. Therefore electricity is produced much of the time in every cycle, not just when the two magnetic poles sweep by the wire.

Furthermore, the size of the magnetic field changes with cycle position (figures 2.5 and 2.6). The magnetic field is indeed strongest when either magnetic pole is closest to the wire. The practical result is that more electrical current is produced when either the North Pole or the South Pole are directly perpendicular to our wire. Yet some current is produced much of the time in every cycle.

However, the following truths from our simplified generator remain: 1) the greatest amount of electricity is made ("induced") when the North or South poles of the magnet are directly over the wire; 2) the direction of electron flow goes one way, then the other, in every cycle of the magnet; and 3) there are also two positions in every cycle where no current is produced at all.

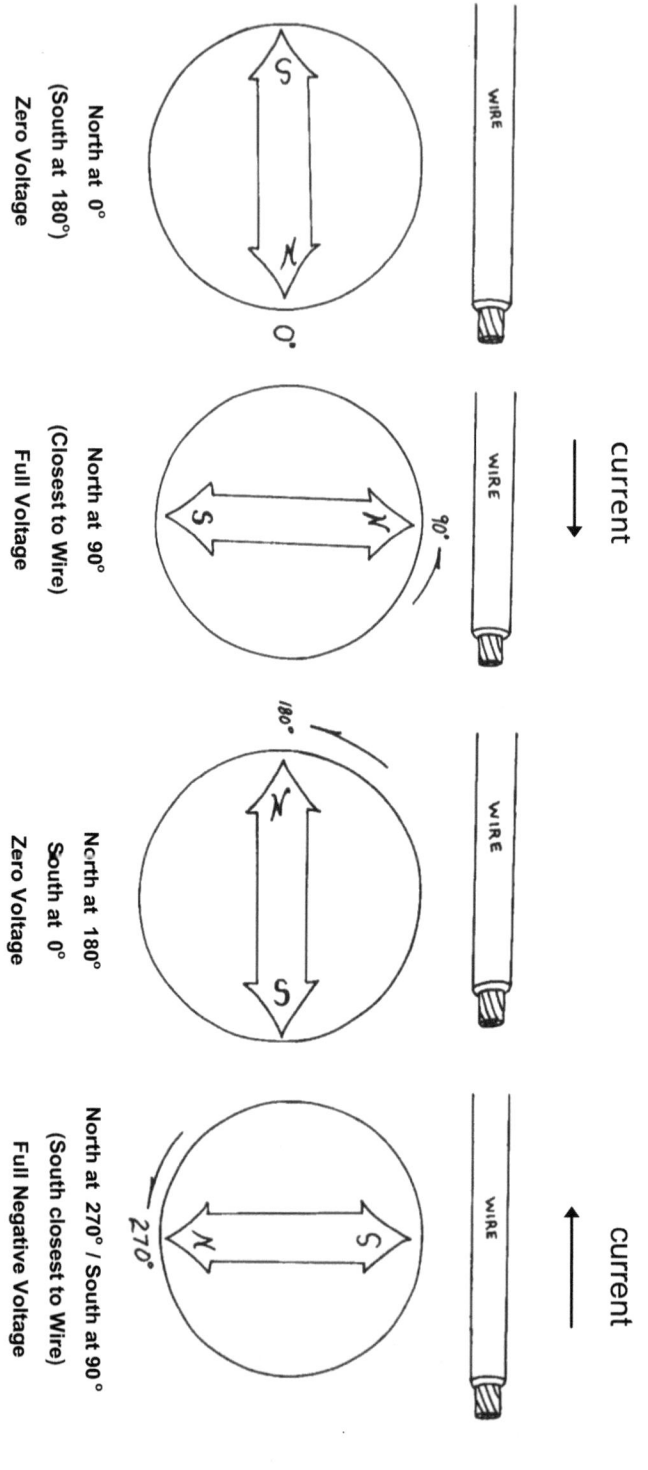

Figure 2.5
Generator in action

Graphing Alternating Current: The Sine Wave

When you see graphs of alternating current, you will see a "sine wave." (Fig. 2.6) In these graphs we are measuring the voltage of the electricity produced for each degree rotation of the magnet. We plot voltage (on the Y axis) versus cycle position (on the X axis).

The cycle position is the position of the "North" end of the magnet, and is usually measured in degrees. The graph is a continuous wave, called a "sine wave." This sine wave comes from 1) the cyclical rotation of the magnet and 2) from the creation of alternating current.

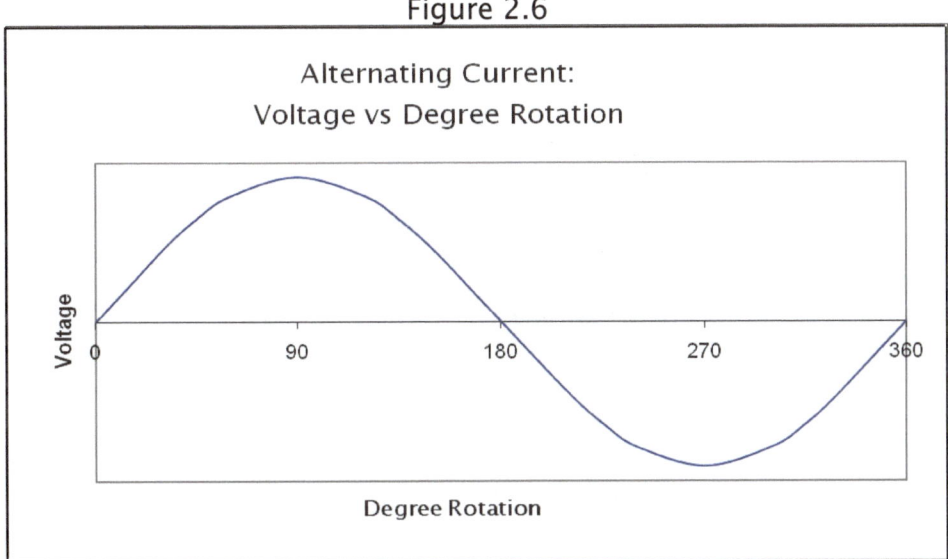

Figure 2.6

The sine wave can be studied using the principles of trigonometry. We will also learn how to interpret these important graphs in chapter 5 of this book.

Three Phase Electrical Supply

Introduction (Fig. 2.7)

A three phase electrical supply involves three wires, rather than just one. The three-phase configuration is found to be the most efficient use of a generator. In general, more than one wire can be added to our set-up. One wire in the generator, as in our original plan above, is called "one phase." A generator with two wires is called "two phase", and a generator with three wires is called "three phase."

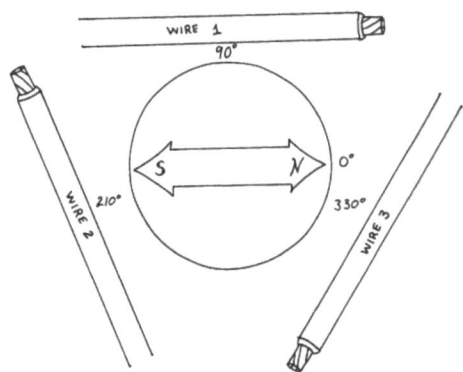

Figure 2.7: Simplified 3-Phase Generator

How Three Phase Supply is Created (Figures 2.8 and 2.9)

In a generator, the three phase electrical supply works as follows. We have one magnet, which rotates. We then have three wires, which are placed around the area of the magnet. An easy way to place these three wires is roughly in the shape of a triangle. This triangle of wires is placed around the area swept by the magnet.

Each time the end of the magnet passes any wire, electric current is produced. This is the process of induction. With three wires, the magnet passes three distinct wires in each cycle, and thus produces three individual currents. The three wires placed around the magnet allow us to create three currents per one cycle of the magnet. This is a 3-for-1 deal, and is thus very efficient.

The phase aspect comes in because each creation of current starts just a bit later than the previous creation. (As soon as the magnet passes a wire, then that phase of electricity starts being produced). We create three different currents, starting at three different times, and yet everything works out smoothly.

Figure 2.8: Creation of 3-Phase Electricity

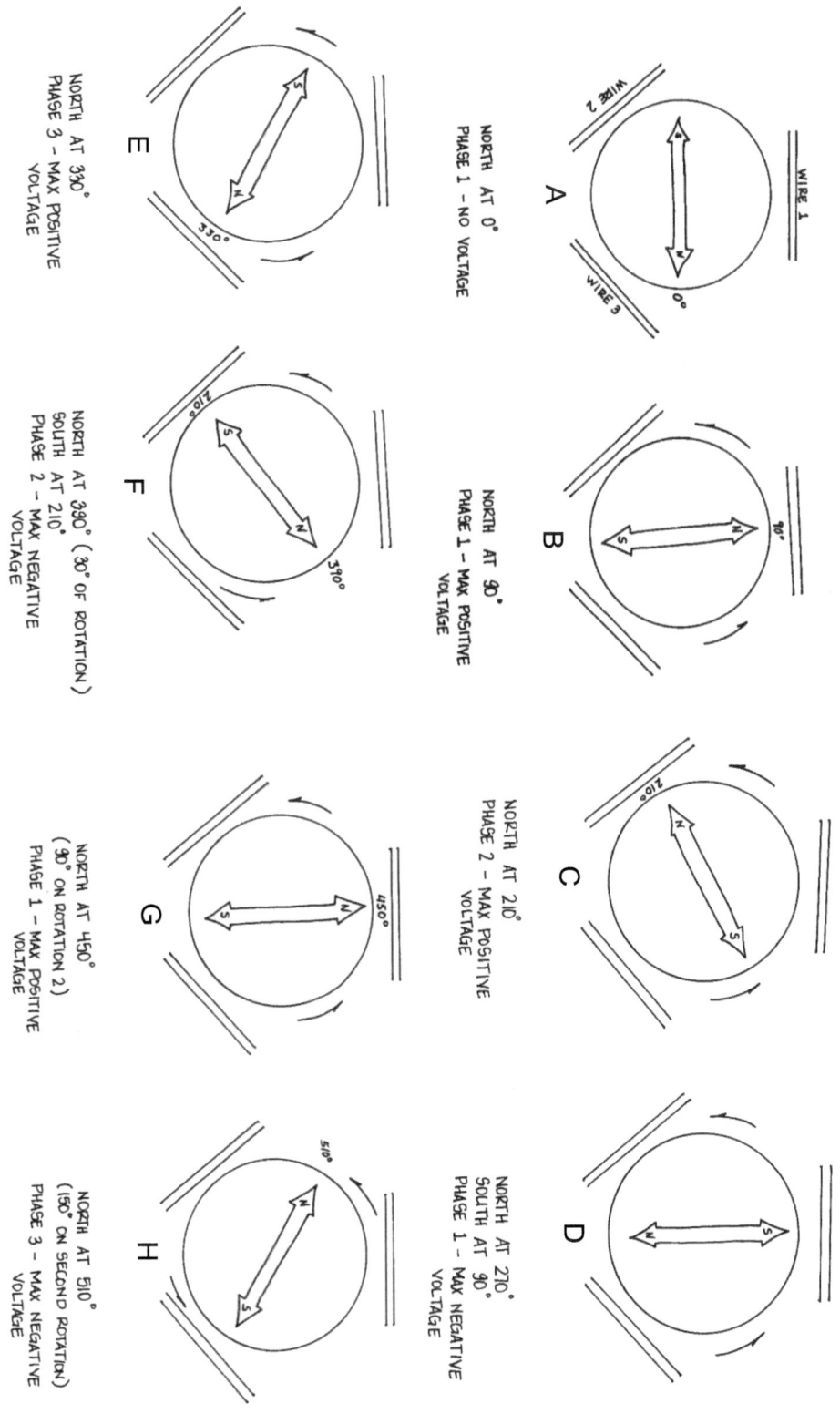

Sine Waves of 3-Phase Electricity (Figure 2.9)

The voltage of a 3-phase electrical supply can be graphed as sine waves in a similar way as the 1-phase electrical supply. As before, we graph voltage versus cycle position. However, in this case, we have three sine waves rather than just one. Each sine wave starts 120 degrees later than the previous one. (A circle is 360 degrees; 360 degrees divided by 3 = 120 degrees.)

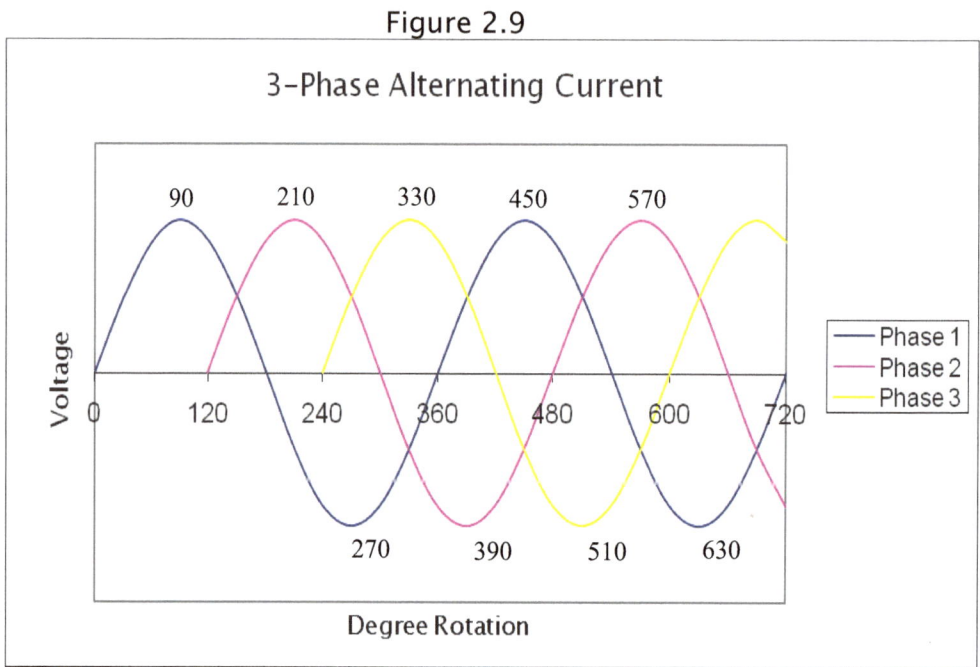

Three Phase Electricity Transmitted Through Power Lines

Most generators produce three phases of electricity, in the method described above. These three phases of electrical power are sent through three separate power lines. At this point, each phase of electrical power is distributed independently across the miles, from the power plant to the individual homes.

Chapter Summary

1. The general purpose of a turbine is to turn the flow of something, such as steam or wind, into rotational movement. The rotational movement of a turbine is required to operate the generator.

2. The flow of molecules provides the initial energy. The turbine converts the energy of flowing molecules into mechanical energy. Then the generator converts the mechanical energy into electrical current.

3. A turbine or generator with a better design will convert one form of energy into the other form more efficiently.

4. Steam turbines and gas turbines require a conversion process. Water and wind turbines do not require any conversion.

5. In a steam turbine, stored energy is released from the fuel. This energy will boil water into steam; the steam then pushes the turbine blades.

6. In a gas turbine, gas phase molecules push the turbine blades. These gas phase molecules are usually created by burning hydrocarbon fuels.

7. The dual turbine combines the gas turbine and the steam turbine in sequence, in order to get the greatest turbine efficiency.

8. Electricity is created by the relative movement between magnets and wires.

9. In a generator, a magnet rotates past a wire. This movement "induces" a current.

10. Alternating current is created by rotating the magnet continuously, so that the opposite ends of the magnet will continuously induce current in opposite directions.

11. Alternating current can be graphed using sine waves.

12. A 3-phase generator is most common. This generator uses 3 wires around the magnet rather than just one wire. A 3-phase generator is the most efficient generator.

13. An alternate form of a generator has wires which rotate and the magnet is stationary. In this generator, the rotation of the wires past the stationary magnet induces a current.

14. Electricity is produced in three distinct phases, and is carried over three distinct power lines from the power plant to your home.

1.3
Voltage, Current, and Power

Introduction
In this chapter we will look at concepts such as current, voltage, and electrical power on the atomic scale. In this chapter you will learn many fundamental concepts on the nature of electricity, including some concepts never before presented.

List of topics for this chapter
1. Electrical Current
2. Voltage
3. Electrical Power
4. Technical Discussions on Current
5. Technical Discussions on Voltage

Electrical Current

Introduction
   Electrical current is usually described as the flow of electrons. Although this is a very convenient way of looking at it, in power lines this is not strictly accurate. When electrical current is transmitted over wires, the electrons do not really travel along the miles of wire. More accurately we can say that electrons bump into each other. In electrical wires, what really constitutes an electric current is the process of consecutive electrons bumping.
   Electrons on a wire do not really flow like water. Electrons are usually fixed on the wire. Furthermore, in a wire the electrons are fixed close together. Therefore these electrons can easily bump into each other. When we push one electron, this electron will push the next electron. This pushing continues through millions and billions of electrons. The net result is a "current."

Dominoes and Direct Current

Dominoes provide a good analogy for direct current. In this analogy we are referring to the common pastime of lining up dominoes, then watching them fall in succession.

We line up the dominoes close together, then we push the first one. The first domino will push the second, the second will push the third, the third will push the fourth, and the process continues down the line. Therefore by pushing one domino we can eventually push all the others.

Imagine the dominoes are electrons. Just as with the dominoes, the electrons are close together. We push the first electron. Then the first electron bumps into the second electron, the second electron bumps the third, the third electron will bump the fourth, and the process continues down the line. Therefore, by pushing just one electron we can eventually push all the others.

Notice two separate events: 1) the bump, and 2) the flow of the bumping. When we watch dominoes fall we will see a flow. Note that this flow is not the flow of dominoes, but rather it is the flow of dominoes hitting each other. The fallen dominoes lay on the ground, they do not actually move. Instead, what "moves" is the flow of dominoes hitting each other down the line.

Similarly, when electrons bump you will also see a flow. However, we are not seeing the flow of electrons. Rather, we are seeing the flow of consecutive electrons bumping. The electrons remain at their positions on the wire, yet the flow of electrons bumping continues along that wire for many miles.

Therefore, we can see that electrical current on the atomic scale is in fact a series of consecutive electrons bumping into each other.

Alternating Current and Dominoes

Imagine that the dominoes are attached to a string. After the dominoes have fallen we can pull the string. The dominoes are then set up just as before, and we can push the dominoes in sequence again.

Alternating current creates the same effect. After the electrons have been pushed one direction, we pull the electrons in the opposite direction. At this point, each electron along the wire has the same position and energy as it did at the start. We can once again push the first electron, which starts the sequence of consecutive electron bumps down the wire.

Therefore, in alternating current we first push the electrons in one direction, then we pull the electrons back to their original state, at which point we may give the electrons another push. This process continues repeatedly and quickly. The net result is a continuous production of alternating current.

Measuring Current

As stated above, electric current is really the flow of consecutive electrons bumping. Therefore, measuring current is simply measuring the number of bumping electrons per time. When we measure current we pick a specific length of power line, then we measure the number of electron collisions over a period of time at that section of wire.

The unit for measuring current is the Amp. We discussed the Amp earlier, but now we can modify our definition for power lines: 1 Amp is approximately $6.5 \times 10^{+18}$ electrons being pushed or pulled per second.

Wire size, Amps, and Current

Wire size determines the maximum amount of current that can travel through the wire.

First, it is physically impossible to conduct more current through a wire than there are electrons. Each wire has a fixed number of electrons, and when all those electrons are bumping into each other it is physically impossible for any more bumping to occur.

Second, only some of those electrons are effective as conductors. Some electrons are not positioned well enough for effective bumping. This means that there is a maximum limit to how many electrons can be bumped per second.

Therefore due to both of those reasons, the wire size determines the maximum amount of electrical current allowed to travel.

Voltage

Voltage can be understood very simply: voltage is energy. More specifically, voltage is the overall energy of many electrons at one location.

Every electron vibrates with some amount of energy. The vibrational energy of that electron is the voltage of that electron.

However, in practical terms we measure the energy of not just one electron but rather we measure the energy of many, many electrons. Therefore in practical applications the term "voltage" means the overall energy of trillions of electrons at one specific location.

The unit for measuring Voltage is called the Volt. The size of 1 Volt is equivalent to 1 Joule of energy per $6 \times 10^{+18}$ electrons.

When we measure voltage, we are really measuring the overall energy of many electrons at one location. We pick a spot on the wire, then we measure the energy of the electrons at that spot. Remember that there is not just one electron at that location but trillions of electrons at that location. Therefore, when we measure the voltage what we are really measuring is the overall energy of trillions of electrons at that one location.

A low number of Volts means that this group of trillions of electrons has low overall energy. Conversely, a high number of Volts means that this group of trillions of electrons has high overall energy.

Electrical Power

Electrical power is combination of the energy of the electrons (the Voltage), and the rate at which electrons hit each other (the Current). If the electrons have lots of energy (high Voltage), then that gives us more power. Similarly, if the electrons hit each other at a fast rate (fast Current), then that gives more power.

Ideally, we want the highest of both. We want the electrons to have as much energy as possible. We also want the electrons to bump into each other at the highest rate possible. When we get the highest of both factors, at the same time, then we have the greatest overall power delivered to our homes.

Technical Discussions on Current

Technically, current is measured in terms of charge per second, not number of electrons per second. However, the unit of amps can easily be defined in terms of the number of electrons per second.

One Amp is defined as: 1 Coulomb of Charge per Second. Regarding Coulombs and current: each electron has a charge of 1.6×10^{-19} Coulombs. This means that 1 Coulomb of Charge is approximately $6.5 \times 10^{+18}$ electrons. Therefore, 1 Amp can be defined as $6.5 \times 10^{+18}$ electrons traveling through a wire per second.

Also, now that you know that electrons in a power line don't actually flow, but rather bump, and that the flow we see is actually the flow of electrons bumping, then we can say that 1 Amp can be defined as $6.5 \times 10^{+18}$ electrons bumping into each other per second.

Technical Discussions on Voltage

Voltage is such a fundamental concept in electrical power that it is worthwhile to have a few technical discussions.

1. Technically, voltage is measured in terms of energy per charge, not energy per electron. However, the unit of Volts can easily be defined in terms of energy per group of electrons. The "Volt" is originally defined to be 1 Joule of energy per 1 Coulomb of Charge, which is then 1 Joule of energy per $6 \times 10^{+18}$ electrons.

2. A "Volt" and an "electron-volt" are really both measurements of energy. The "Volt" is the number of Joules of energy for trillions of electrons. The "electron-volt" is the number of Joules of energy for only one electron. The value of the electron-volt can be easily figured from those values above. The resulting value is 1 eV = 1.6×10^{-19} Joules of energy for 1 electron. (You will also note that the number "1.6×10^{-19}" is the same number that is in the definition of Coulombs).

3. Earlier in this book we discussed the electron-volt, and related the eV to number of Joules per one electron. However, you will note that in our list of energy units that "electron-volt" is on the list, but not "Volts". Technically, we could list Volts as a unit of energy, just as we did for eV, but in practical terms no one does this.

Also, we could create conversions for Volts into other units of energy. However, again, no one does this. The way in which people work with voltage on a practical level just doesn't bring up the need to discuss other units of energy at the same time.

Summary

1. Electrons on a wire do not usually flow. Electrons are usually fixed on a wire.

2. Electrical current is really the flow of consecutive bumping electrons.

3. The act of measuring electrical current is really measuring the number of electrons bumping into each other per second.

4. Each wire has a fixed number of electrons. Therefore, the wire size determines the maximum possible electrical current. It is physically impossible to conduct more current through a wire than there are electrons available for bumping.

5. Voltage is the overall energy of many electrons at one location.

6. The Volt is the unit of measurement for the overall energy of many electrons at one location. 1 Volt = 1 Joule of energy per $6 \times 10^{+18}$ electrons.

7. Electrical Power is combination of Energy of the electrons (Voltage), and the Rate at which electrons hit each other (Current). When we get the highest of both factors, at the same time, then we have the greatest overall power delivered to our homes.

1.4
Alternating Power, Frequency and Hertz

Introduction
Electrical power is created as alternating power. Electrical power is sent forward, then in reverse, repeatedly. The frequency of this alternating power is very important. Deviations in frequency or bursts of high frequency can seriously damage electrical equipment.

List of topics for this chapter
1. Alternating Power Basics
2. Frequency Basics
3. Alternating Voltage and Alternating Current
4. Forward Power vs. Alternating Power

Alternating Power Basics

Introduction
　　Alternating power is created by the magnet in the generator. The movement of the magnet past the wire creates a burst of power flowing down the power lines. Then the opposite pole of the magnet pulls the power back (toward the generator). This action is done repeatedly, and very quickly, thereby creating continuous alternating power.

Three Phases and Frequency
　　There are always three phases of electricity created by the same generator. Therefore, all three phases of electricity coming from the same generator have the same frequency.
　　However, once created the three phases of electricity are independent. Each phase runs on a separate wire all the way from the power plant to your home. Along the way many things can happen. Therefore, anywhere on the path one phase may have frequency problems though the other two phases have perfect frequencies. Therefore all three phases need to be monitored for frequency.

Frequency Basics

Introduction

Frequency is a fundamental factor in electrical power. Theoretically, the creation of alternating power could be done at any frequency. However, electrical systems must operate at very close to the same frequency in order to function. For example, power plants which are connected must use the same frequency. Also, machines connected together in the same manufacturing facility must operate on the same frequency. Furthermore, electrical equipment at substations can become permanently damaged by changes in frequency. Therefore, we must have a uniform standard of frequency, and that standard must be adhered to strictly.

Important Terms Related to Frequency

cycle: Each time the power is sent forward and then pulled back is considered to be one cycle. This cycle is graphed as a sine wave.

frequency: We can measure how fast the power is sent forward and pulled back. In general terms, this rate is "the number of cycles per time." The number of cycles per time is called the frequency.

The frequency of alternating power is determined by the frequency of the rotating magnet inside the generator.

Hertz, Hz: The frequency of alternating power is measured in the unit of the Hertz. A Hertz is defined as the number of cycles completed per second. In America, 60 cycles per second (aka 60 Hertz) has been designated as the common frequency for all electrical systems.

Alternating Voltage and Alternating Current

Alternating power is a combination of alternating voltage and alternating current. Ideally, both the voltage and current should have the same frequency and both should be in phase. Therefore, in many cases only one factor (usually Voltage) is tracked. However, the current can deviate from voltage. (Details will be mentioned in throughout this series of books). Therefore, both current and voltage should be tracked at all times.

Therefore, we will have two factors to track. We will be creating alternating voltage, which can be graphed on a sine wave, and should be a frequency of 60 Hertz. Similarly, we will be creating alternating *current*, which can be graphed on a sine wave, and should also be a frequency of 60 Hertz.

Alternating Power and Forward Power

How can current be both alternating and move forward to our homes? Similarly, how can voltage be alternating and yet move forward? First, remember our analogy of the falling dominoes. Once the initial push has been started, the flow continues for a long time. The flow of electrons bumping will continue down the line long after the first electrons are pulled back toward the generator.

Second, the current (flow of electrons bumping) occurs at very high speeds; one burst of electrical energy can travel many miles in the blink of an eye. For example, in a long-distance high voltage power line, an electrical current can travel up to 200,000 km/second.

Perhaps most important, the time between bursts of electrical energy is very short. Referring to the dominos, time is required to pull the dominoes back up. It is no different for electrons, however this process happens very, very fast. The generator sends a forward burst of electricity in 1/120 of a second. Similarly, the generators reset everything in another 1/120 of a second. Therefore, we never notice any gap in time. To us, it appears that we get a constant stream of electrical energy.

Chapter Summary

1. The creation of alternating power is a continuous cycle. The frequency of the cycle is measured in Hertz.

2. Hertz is defined as the number of cycles completed per second.

3. The frequency of the cycle is determined by the rotational speed of the electrical generator.

4. Theoretically, the creation of alternating power could be done at any frequency. However, electrical systems must operate at very close to the same frequency in order to function properly.

5. In America, 60 cycles per second (60 Hertz) has been designated as the common frequency for all electrical systems.

6. Electrical Power is a combination of Voltage and Current. Therefore, where we have alternating power we will have both alternating voltage and alternating current. We must track both factors.

7. All three phases of electricity coming from the same generator have the same frequency.

8. After the electricity is created, the three phases of electricity are independent. Along the way many things can happen which can affect the frequency on any one of the lines.

1.5
Understanding the Current and Voltage Sine Wave Graphs

Introduction

We have said many times that electrical power is a combination of Voltage and Current. These are the two primary factors of electricity. As such, they are very often measured, graphed, and studied. Therefore, it is worthwhile to become familiar with each sine wave graph.

Whenever a graph of voltage or current comes to our attention, we should have the ability to examine that graph and understand the specific situation. When we are able to read and understand these graphs we will be able to make effective decisions. In order to have this ability we should understand what is really happening, not just mathematically, but what is really happening along the wires.

In this chapter we will look at the voltage sine wave graph and we will look at the current sine wave graph from the level of electrons. By the end of this chapter we will learn how to read these very important graphs.

List of topics for this chapter
1. Voltage and Current can Each be Graphed as Sine Waves
2. Understanding Voltage Sine Wave Graphs
3. Understanding Current Sine Wave Graphs

Voltage and Current can each be Graphed by Sine Waves

When we create alternating current there are in fact two factors: voltage and current. Most of the time we track only the voltage, however we can track the current as well. Both voltage and current are created in the same way (by the rotation of the magnet in the generator). Therefore both can be graphed as sine waves. (Figures 5.1 & 5.2)

Notice that each graph is examining one section on the power line. For example, Figure 5.1 measures the voltage, over time, along one piece of the power line. Similarly, Figure 5.2 measures the current, over time, along that same piece of the power line.

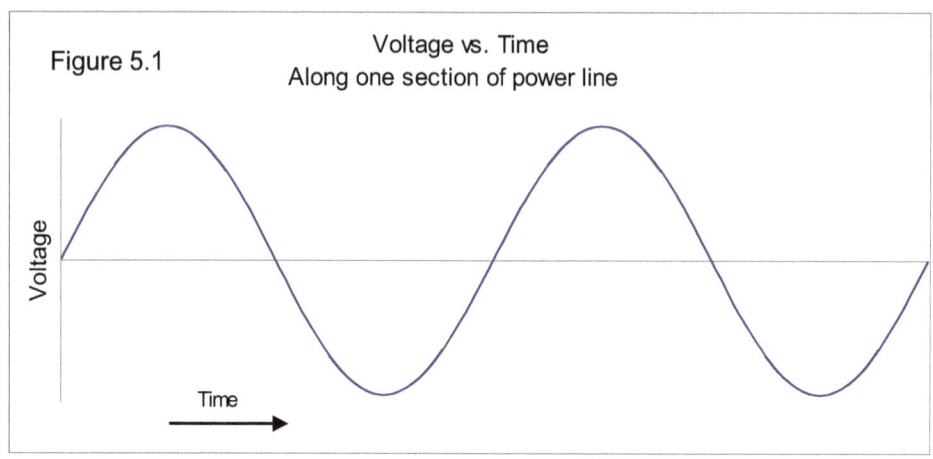

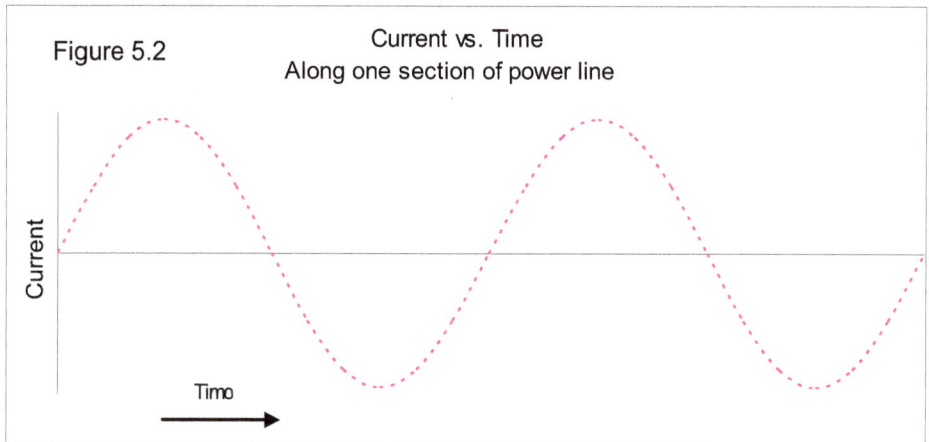

Understanding Voltage Sine Wave Graphs

Introduction

We will now look at the voltage. We will use graph 5.1 above, but labeled at various stages. (This is now Figure 5.3.) The voltage sine wave is a graph which tracks Voltage vs. Time, at a specific section of power line. On the atomic scale, this means that the sine wave tracks:

    a. The overall energy of electrons at a specific location
    b. The direction of energy transfer

Focusing on One Small Section of Power Line

When we use these graphs, we are focusing on just one small section of a power line. We don't think too much about what comes before, nor do we think too much about what comes after. We are only looking at what the electrons do at a particular section of the line. This is a fundamental concept to remember as you observe these graphs.

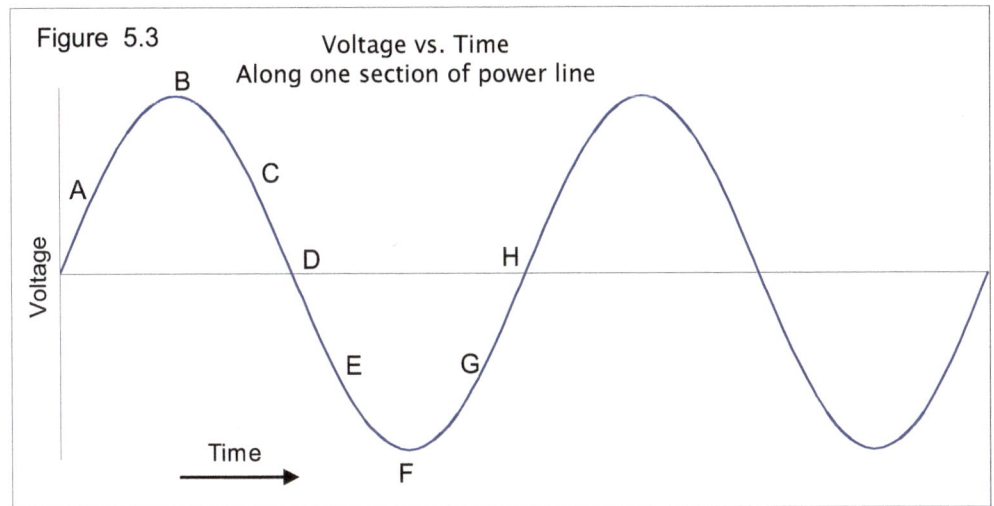

Figure 5.3 Voltage vs. Time Along one section of power line

Peaks and Dips

The peaks and the dips in the graph tell us which direction the energy flows. The sections of the graph with the peaks are tracking the period of time when the energy is moving forward, from the generator to the home. The sections of the graphs with the dips are tracking the period of time when the energy is flowing backward (toward the generator).

What occurs at each stage

The stages of the Voltage vs. Time graph are as follows:
• To start, the electrons in this region have their normal level of energy.

• At point A the energy is moving forward, from the generator to the homes. The overall energy of the electrons at this location of wire begins to increase.

• At point B the overall energy of the electrons is the maximum it will be.

• At point C the overall energy of the electrons in this region begins to decrease.

• At point D all the electrons in this region once again have their normal level of energy.

• At point E the energy starts to flow backwards. The overall energy of the electrons at this location begins to increase. (Backwards flow; increase)

• At point F the overall energy of the electrons in this region is the highest it will be. This is similar to point B. However, the dip reminds us that the energy is flowing in reverse.

• At point G the overall energy of the electrons in this region decreases. The energy is still flowing backwards.

• At point H the overall energy of the electrons in this region is again at their normal energy level. From this point, the process repeats.

Understanding Current Sine Wave Graphs

Introduction

We will now look at the current. We will use graph 5.2 above, but labeled at various stages. (This is now Figure 5.4). The sine wave for electrical current is a graph which tracks Current vs. Time, at a specific section of power line. On the atomic scale, this means that the sine wave tracks:

a. The overall rate at which electrons bump at a specific location.

b. The direction of the consecutive electrons bumping.

Focusing on One Small Section of Power Line

When we use these graphs, we are focusing on just one small section of a power line. We don't think too much about what comes before, nor do we think too much about what comes after. We are only looking at what the electrons do at a particular section of the line.

Peaks and Dips

The peaks and the dips in the graph tell us which direction the current flows. The sections of the graph with the peaks are tracking the period of time when the current is moving forward, from the generator to the home. The sections of the graphs with the dips are tracking the period of time when the current is flowing backward (toward the generator).

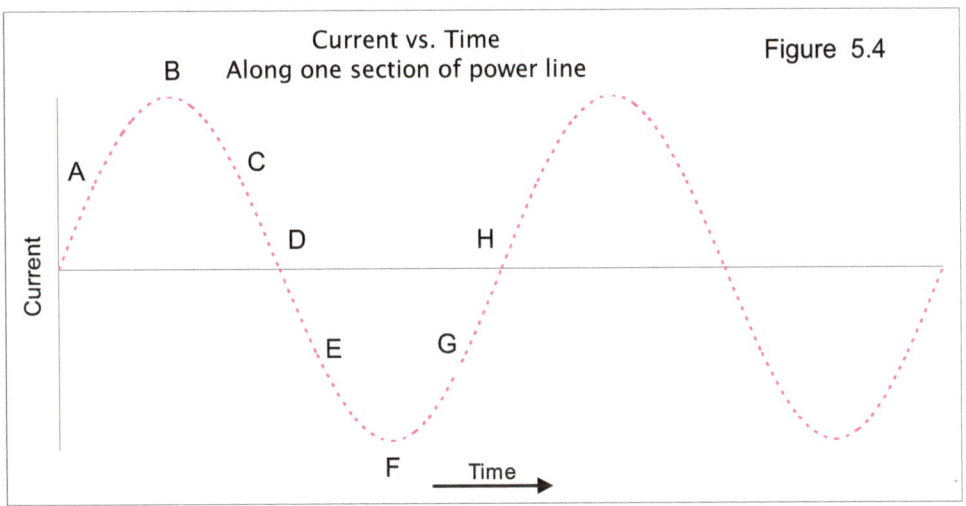

Figure 5.4

What occurs at each stage

• To start, there is no current flowing at this location. Electrons are not bumping into each other at this time.

• At point A, the current (the rate of consecutive electrons bumping) at this location of wire begins to increase. The current is moving forward, from generator to the homes.

• At point B, the current is the maximum it will be.

• At point C, the current in this region begins to decrease.

• At point D, no current flowing at this location.

• At point E, the current is flowing backwards, toward the generator. The current at this location begins to increase.

• At point F, the rate of consecutive electrons bumping in this region is again the highest it will ever be. However, the dip reminds us that the current is flowing backward.

• At point G, the current in this region decreases. The current is still flowing backwards.

• At point H, there is no current. From this point, the process repeats.

Summary

1. Voltage and Current can each be graphed by sine waves.

2. The voltage sine wave is a graph which tracks Voltage vs. Time, at a specific section of power line. On the atomic scale, this means that the voltage sine wave tracks:
    a. The overall energy of electrons at a specific location
    b. The direction of energy transfer

3. The peak and the dip in the Voltage vs. Time graph tell us the direction which the energy flows. The section with the peak is tracking the time when the energy is moving forward, from the generator to the home. The section with the dip is tracking when the energy is flowing backward, toward the generator.

4. The sine wave for electrical current is a graph which tracks Current vs. Time, at a specific section of power line. On the atomic scale, this means that the current sine wave tracks:
    a. The overall rate at which electrons bump.
    b. The direction of the consecutive electrons bumping.

5. The peak and the dip in the Current vs. Time graph tell us the direction which the current flows. The section with the peak is tracking the time when the current is moving forward, from the generator to the home. The section with the dip is tracking when the current is flowing backward, toward the generator.

1.6
Batteries

Introduction

Batteries are capable of storing electrical energy, which can later be used to produce electrical power. There are numerous types of batteries and many aspects of batteries which can be studied. However our primary interest in batteries is in their applications and connections with electrical power systems. Therefore this chapter will discuss concepts of batteries that will be important to other electrical power topics in this series of books.

Note that batteries will be discussed in greater detail in a subsequent book in the series. Therefore this chapter will provide an overview of batteries, and will describe the most important concepts of batteries. Additional details will be found in the volume dedicated to Batteries and Corrosion.

List of topics for this chapter
1. Battery Power versus Turbine-Generator Power
2. Overview of How Batteries Work
3. Anode and Cathode Overview
4. Electrolyte Solution to Create Free Electrons
5. Pull Force of Nuclei and Voltage of Batteries
6. Corrosion as Related to Anode and Cathode
7. Life Cycle of Batteries
8. Rechargeable vs. Not Rechargeable
9. Specific Differences between Battery and Generator
10. Battery Use for Solar Power
11. Size of Batteries
12. Depth of Discharge Overview
13. Depth of Discharge in Detail

Battery Power versus Turbine-Generator Power

Before we begin discussing batteries it is very important that the reader understand one fundamental concept: batteries operate very differently from the turbine-generator system we have been discussing.

In general, most of the books in this series discuss technologies which are based on the mechanisms of the turbine-generator system discussed previously. However, for this chapter the reader must set all those other mechanisms aside.

The differences between batteries and the turbine-generator system are so vast that the reader must consider batteries as a totally separate entity from all other electrical systems. In particular, the mechanisms of current and voltage are very different in the two systems. Consequently, all operations related to the battery are completely different from the operations of the turbine-generator system.

Therefore, we will approach this chapter on batteries with the perspective and understanding that the mechanisms of batteries are very different from anything discussed previously. In other words, set aside what you have read in last four chapters, and open your mind to thinking in a totally new way. Doing this now will help you understand both the batteries and the turbine-generator systems more clearly.

Ready? Now we may begin.

Figure 6.1: Basic Battery Design

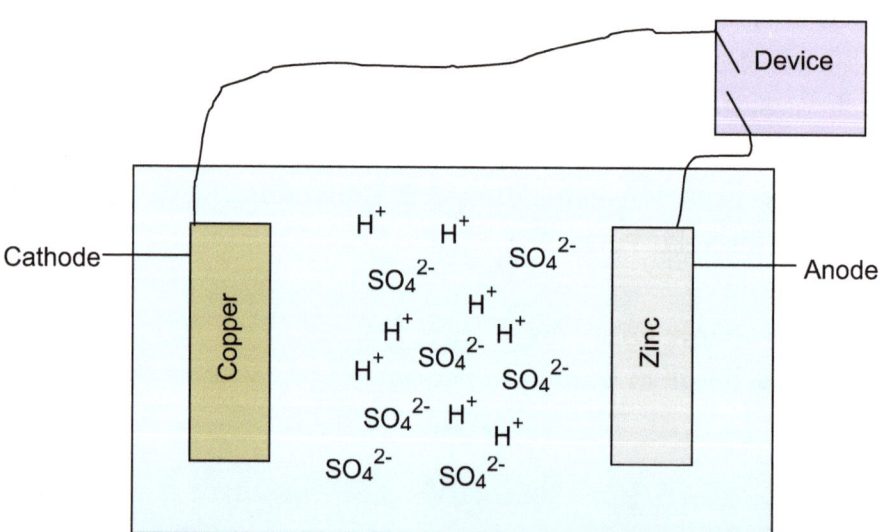

Overview of How Batteries Work

In order to create a battery we need three items: 1) two metal bars, each made of a different type of metal; 2) an electrolyte solution (usually a strong acid or very reactive negative ions); and 3) a wire connecting the two metal bars. (Figure 6.1)

The basic process of the battery is as follows:

1. The electrolyte solution reacts with the first metal (the anode). This reaction creates free electrons.

Figure 6.2: Reaction at Anode, Creating Free Electrons

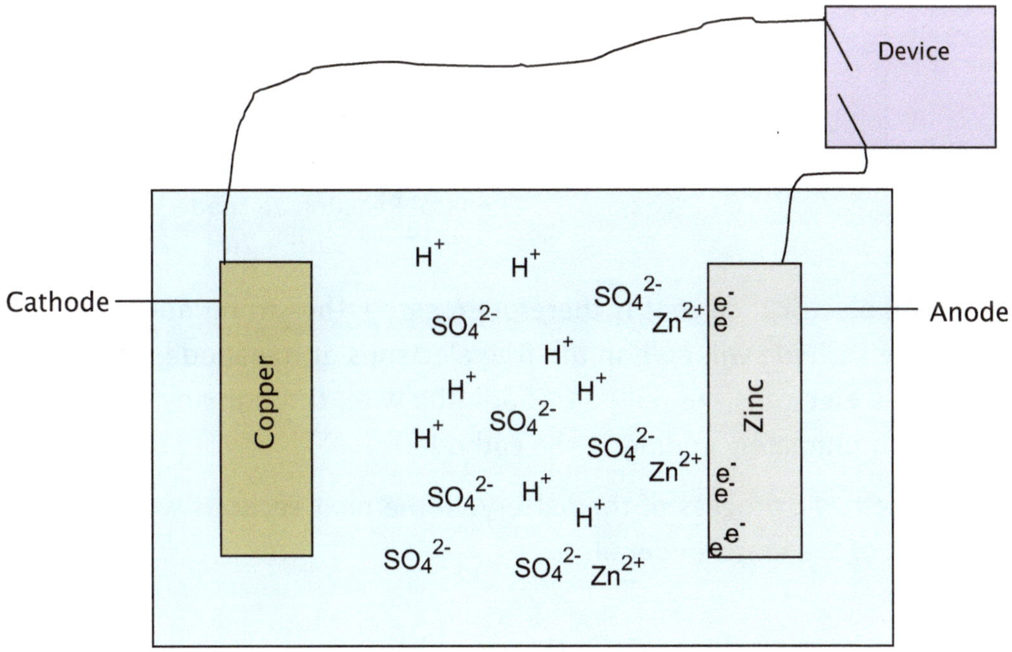

Chemical Reaction at Anode:
$$SO_4^{2-} + Zn \longrightarrow SO_4^{2-} + Zn^{2+} + 2e-$$

2. The nuclei of the second metal (the cathode) has a very strong pull, similar to a gravitational pull, on any free electrons nearby. Therefore this second metal will naturally pull any free electrons which were created at the first metal.

3. A wire is connected between the two metals which allows the free electrons to travel from one metal to the other metal more easily.

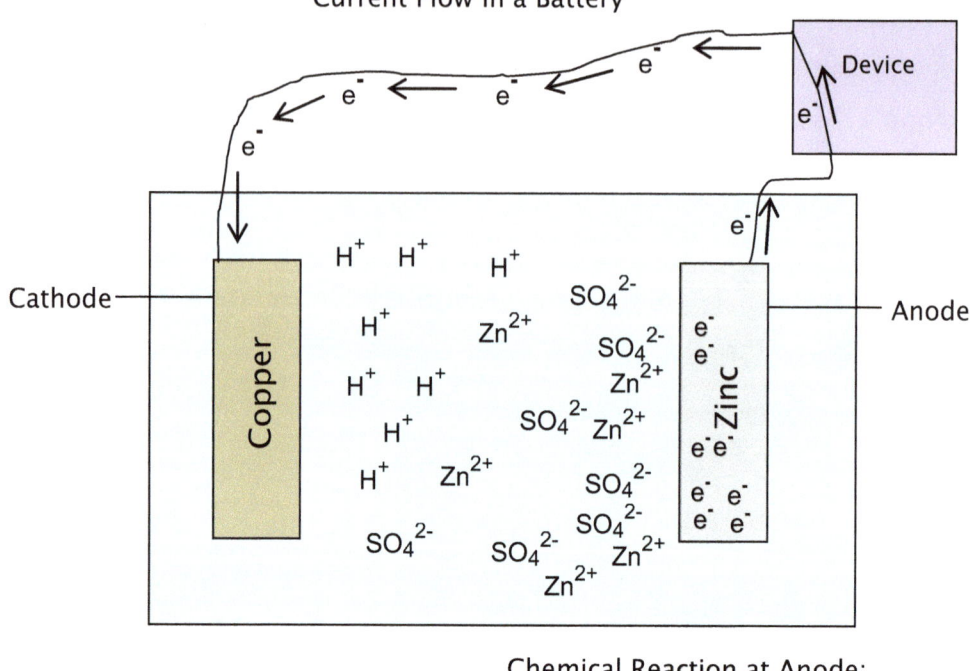

Figure 6.3: Process of Current Flow in a Battery

Chemical Reaction at Anode:
$$SO_4^{2-} + Zn \longrightarrow SO_4^{2-} + Zn^{2+} + 2e-$$

4. An electrical current is therefore created: the strong nuclei at the cathode will pull on the free electrons at the anode. These free electrons are pulled through the wire, through any device, and ultimately ending at the cathode.

This is the basic process of the battery. In the next sections we will discuss each of the steps in detail.

Anode and Cathode: Overview

The two metals are called the "anode" and the "cathode." Electrons always flow through the wire from the anode to the cathode. A metal is only an anode or cathode in respect to the other. These categories are primarily due to the pull strength of each type of atom on nearby electrons. (See the section on Pull Strength below). For our examples, we use Zinc as the anode, and Copper as the cathode.

The anode is the metal which creates free electrons, and then loses those electrons in the circuit. Technically we say that the anode has been "oxidized." The cathode is the metal which gains electrons in the circuit. Technically we say that the cathode has been "reduced."

Electrolyte Solution to Create Free Electrons

The first step in the process is to create free electrons. We do this by using an electrolyte (often a strong acid) to separate an atom from its electrons.

One common electrolyte solution is sulfuric acid (H_2SO_4). Sulfuric acid exists in water as very loose ions of H^+ and SO_4^-. The negative SO_4^- ion attacks the metal. In our example the SO_4^- ion attacks the zinc. The ion is powerful, and rips the zinc atom away...not only from the zinc bar but from two of the electrons as well. The net reaction is: $Zn \rightarrow Zn^{2+} + 2\ e^-$.

Now we have 2 free electrons. These electrons can be used in our electrical current.

Pull Force of Nuclei and Voltage of Battery

The relative pull strengths of each atom in the battery are what creates the current and the voltage in our battery. In this section we will provide a brief overview. (For more details see the book on Batteries).

The nucleus of each atom has a force similar to gravity which pulls on nearby electrons. And just as different planets have different strengths in gravitational pull, so it is with different types of atoms. Specifically, each type of atom has various strengths when pulling on electrons.

Consequently, the direction of current in a battery, and the amount of voltage in a battery, will depend on the relative difference between the pull strength of two different materials.

For example suppose we had one metal with a pull strength of 2 volts, and another metal with a pull strength of 10 volts. Now we put them in our battery.

First, note that we will be removing our free electrons from the metal with 2 volt pull. Why? Because of the weaker pull strength. The metal with the 2 volt pull has a much weaker hold on electrons than the metal with the 10 volt pull, and therefore it is easier to remove the electrons off the metal with the weaker pull strength.

Second, note that the electrons will flow from the metal with the 2 volt pull to the metal with the 10 volt pull. Why? Because the pull force is stronger in the 10 volt pull. Any free electrons will move where they are most strongly pulled. In this case, that is the metal with the 10 volt pull strength.

Now we come to the voltage of the battery as a whole. If the pull strength of the stronger metal is 10 volts, and the pull strength of the weaker metal is 2 volts, then the electrons are being pulled with a net value of 8 volts.

Viewed another way, there is a tug of war where each metal is pulling on the electrons. Who will win? The metal with the greatest strength. Therefore, the metal with the greatest pull strength will dictate the direction of current.

And what is the net force on the electrons which are being pulled? We have 10 volts of energy pulled forward, minus two volts of energy pulled back, which results in a net value of 8 volts of energy on all electrons being pulled forward. Hence we have a value of "8 volts" on our battery.

Corrosion as Related to Anode and Cathode

When you look at a battery after it has been in operation for some time, you will notice that one metal has progressively been worn away. You will also notice that the other metal has become fuzzy. (Fig 6.4)

The anode is the metal with the hole. This hole exists because many positive ions from the anode have left the metal and traveled through the solution. Similarly, the cathode is the metal which grows fuzzy. The fuzz on the cathode is the same multitude of positive ions which left the anode.

For example, at the zinc anode the reaction takes place which creates free electrons, as well as the ion Zn^{2+}. Each time the reaction takes place a Zinc atom is removed. You will see a hole grow. Eventually the Zinc bar will be disintegrated.

Then on the other side of the battery we see that the Zn^{2+} ions have migrated. The free electrons which have gathered at the cathode will attract the Zn^{2+} ions. The Zn^{2+} ions travel to the cathode and latch onto the outside. This is what gives the cathode its fuzzy appearance.

You can easily remember which metal does what by this simple memory device: Anode, Oxidized, Eaten Away. (All of these words begin with vowels). Then for the other electrode: Cathode, Reduced, Fuzzy. (All of these words begin with consonants).

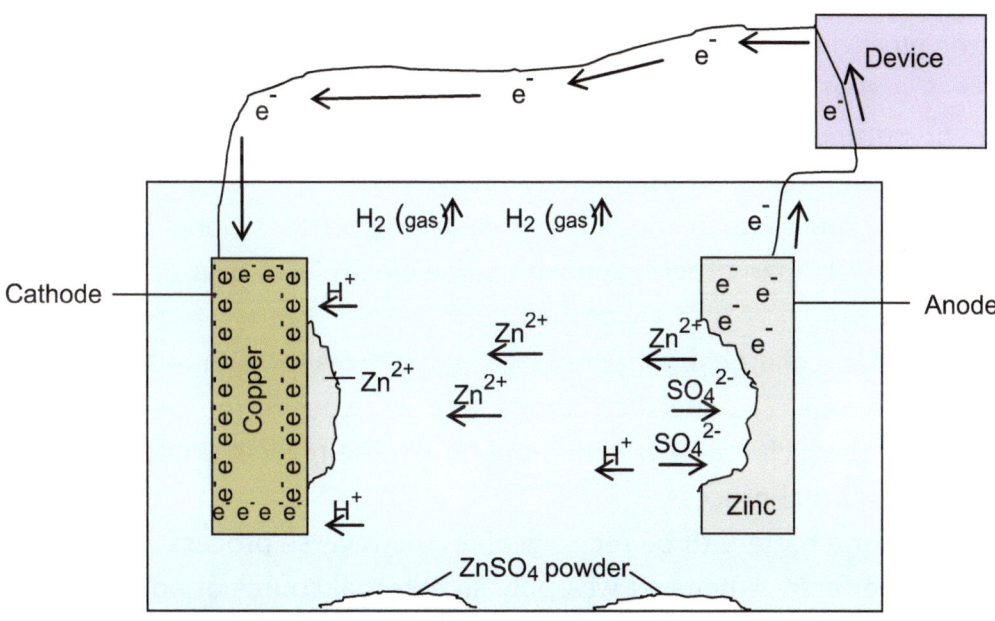

Figure 6.4 Condition of Battery after Operating for Some Time.

Note that this process is a type of corrosion. After this process has been done thousands of times, there is a visible dent in the anode. If left unchecked, there will be a hole. This corrosion is significant not only in batteries, but is also significant in power lines and pipelines.

Life Cycle of Battery

The process described above will create electrical current, as long as there are anode atoms and anode electrons remaining.

Note that the chemical reaction in step one removes a metal atom. Therefore, with each free electron created, some atoms will be removed from the anode. Eventually, all the atoms of the anode will be removed, and all the electrons will have been sent through the wire. When there are no more atoms or electrons at the anode, then the battery cannot produce any more power.

Rechargeable vs. Not Rechargeable

In theory we can recharge any battery. If we apply an external source of power to our depleted battery (specifically at the cathode), then we can push the electrons away from the cathode, through the wire in the reverse direction, and back to the anode. We can therefore rebuild the anode, by putting back electrons, and by taking back the positive metal ions which were ripped off from anode previously.

However, rebuilding an anode is not easy. The process works better with some metals than others, and with some electrolytes over others. Even then, the process is rarely perfect.

Thus, a true rechargeable battery must have the right combination of: 1) a specific anode, 2) a specific cathode, and 3) a specific ionic solution (or ionic paste) which together can 4) operate in the reverse process as easily as the normal process.

In order for a battery to be rechargeable, the reverse process must go simply and perfectly, whenever we apply an external source of power.

Specific Differences between Battery Power and Turbine-Generator Power

Introduction

Earlier we emphasized the concept that the mechanisms of the battery are very different from the mechanisms of the turbine-generator system. At this point it is helpful to note the primary differences between these two power generating systems.

As you work with either system, or as you learn about either system in greater detail, it can be helpful to remember these differences.

1. AC vs. DC
    a. The turbine-generator system creates Alternating Current.
    b. The battery system creates Direct Current.

2. External Energy vs. Internal Energy
    a. The turbine-generator system requires an energy source outside of the generator (such as coal or wind).
    b. The battery system has its own energy supply.

3. <u>Fixed Region vs. Traveling Down the Wire</u>
    a. In the turbine-generator system the electrons remain on a fixed position in the wire.

    b. In batteries the electrons actually move freely down the wire.

4. <u>Electrons Bumping vs. Electrons Flowing</u>
    a. In the turbine-generator system, an electrical current is primarily the flow of consecutive electrons bumping down the wire.

    b. In batteries an electrical current is primarily the flow of free electrons traveling down the wire.

5. <u>Attached to the Atom vs. Free Electron</u>
    a. In the turbine-generator system, electrons can remain attached to the atom and yet create current. This is because current exists primarily through electrons bumping.

    b. In the battery system, we must create free electrons (these are electrons which are separated from the atom) in order to create electrical current. This is because in the battery the current is the flow of free electrons traveling down the wire.

6. <u>Energy Applied by Magnetic Field vs. Pull of Nuclei</u>
    a. In the turbine-generator system, the energy is applied to the electrons by the push or pull from the external magnetic field.

    b. In batteries, the energy is applied to the electrons primarily from the pull of the nuclei in the destination cathode.

7. <u>Amount of Voltage is Related to Magnetic Field vs. Pull Strength</u>
    a. In the turbine-generator system the amount of voltage is directly related to the strength of the applied magnetic field.

    b. In the battery, the amount of voltage is directly related to the relative differences in atomic pull between two different metals.

8. <u>Voltage: Vibration vs. Forward Motion</u>
   a. In both the turbine-generator system and in the battery, voltage is the overall energy of electrons. However, the type of energy which is emphasized is different in each system.

   b. In the turbine-generator system the voltage energy is primarily vibrational energy, with some of the remaining energy in the orbital position and the forward motion.

   c. In the battery the energy is primarily in the forward motion of electrons, with some energy in the vibration.

9. <u>Unlimited Supply vs. Limited Supply of Electrons</u>
   a. In the turbine-generator system electrons are never created or removed from the system, and therefore we have an unlimited supply of electrons. Consequently, in the turbine-generator system we can create power forever using the same set of electrons.

   b. In the battery, electrons are created through chemical reaction, and therefore there is a limit to the supply of electrons. Consequently, in the battery there is always a limited supply of power due to the limited quantity of electrons.

10. <u>Returning Electrons to Original Position: Perfectly vs. Imperfectly</u>
    a. In the turbine generator system we can return electrons to the original position by reversing the direction of the magnet. Returning electrons to the original position works perfectly in the turbine-generator system.

    b. In the battery we return electrons to the original position by applying an external current to the cathode. Returning electrons to the original position is much more difficult and less reliable in the battery.

Battery Use for Solar Power

Introduction

Batteries are most commonly used with solar power. The batteries are charged during the day then the energy is used at night and on cloudy days when the sun is not available.

Batteries are also essential if the application is to be completely electrically independent or "stand-alone". (Stand-alone applications are best for smaller power needs such as lighting and communication devices).

Types of Batteries for Solar and Stand-Alone Applications

There are a variety of batteries available. Lead-acid batteries are the ones most commonly used for solar power and stand-alone applications. (Note that the car battery is a type of lead-acid battery). Types of lead-acid batteries include: lead-antimony; lead-calcium; and the lead-antimony/calcium. A few other common types of batteries (not lead) for stand-alone applications include: nickel-cadmium (also called NiCad); zinc-oxide; and sodium-sulfur.

Charging the Batteries

Using solar power in combination with a battery means that you will be in fact charging (and recharging) the battery. However, remember what we said earlier: charging a battery requires rebuilding an anode, and the process does not always go perfectly. Therefore, batteries for use in solar power, like batteries used for cars, must be able to recharge with near perfection every time. Furthermore, if you want to use a battery for solar power, then the battery must effectively rebuild the anode multiple times. Consequently, solar power requires batteries which are specially designed for effective charging and recharging.

Details of battery charging and solar power will be discussed in a subsequent book. For now we can say the following: the solar array will collect energy from the sun, convert the energy into direct current, then force the electrons in the battery to go in reverse order (from cathode to anode), thereby rebuilding the anode.

Of course, the solar cell can operate the device directly while at the same time recharging the battery. These details will be discussed in subsequent books.

Size of the Batteries

Introduction

In general, the size of the batteries you need will depend on several factors: how much energy you want to store for later, how much power a particular device requires to operate, and how independent you wish to be from the utility company.

Note that when we speak of the "size" of a battery, there are really several concepts. A "larger" battery can come in several forms:
1. A single battery with greater storage capacity
2. Multiple batteries connected together, creating a total, large storage capacity
3. A single battery with a greater depth of discharge

Greater Storage Capacity

In order to obtain a larger storage capacity, you can either increase the size of your electrodes, or choose a different pairing of materials.

If you increase the size of your electrodes, then there are more electrons available to take, and therefore more current which can be created. Larger electrodes can be made simply by having a single electrode of a larger size. However, it is more common to have multiple electrodes (such as several zinc bars) placed next to each other.

You can also increase the storage capacity of your battery by choosing a different combination of anode and cathode. With a different choice of anode and cathode you will have different options for voltage. If the differences in the nuclei pull strength of the anode versus cathode is greater in a different pairing of metals, then your voltage will be greater by using that pairing. This is one of the reasons why you will read so many options in the battery catalogs.

Multiple Batteries Connected Together

Another way to increase battery size is through connecting multiple batteries together. Connecting your batteries together in series (see chapter eight) will allow the energy from each battery to add, and therefore the total energy is the sum of energies in all the batteries.

You can also use multiple batteries as multiple storage units. For example, recharge several at the same time using solar power, yet later use only the specific batteries you need.

Depth of Discharge

Depth of discharge refers to how much current you can draw from the anode before the battery needs to be recharged. The details of Depth of Discharge will be discussed below.

Depth of Discharge: Overview

Introduction

The term "discharge" refers to electricity leaving the battery. The "depth" of discharge refers to how much power of the battery was used before recharging. There are different types of discharge, depending on amount of discharge and how often the discharge occurs before recharging. Battery capabilities and lifetimes are described using these various discharge terms.

I will use the example of the car battery when explaining these terms. This is because a car battery is similar to the batteries for solar cells, and because many people are familiar with the car battery.

Brief Summary of Depth of Discharge Terms

The following is a list of discharge terms, along with a brief explanation. A more detailed explanation is presented later.

1. Shallow discharge: using only small amounts of electricity from the battery before recharging.

2. Deep discharge: using a large amount of electricity from the battery before recharging.

3. Self-discharge: slow drain from the battery even though the battery is not actually being used.

4. Daily depth of discharge: the amount of electricity that we can take from the battery on a daily basis, assuming that we recharge the battery soon after.

5. Maximum allowable depth of discharge: the maximum power you can take from the battery without reaching deep discharge amounts.

Depth of Discharge: Details

Shallow Discharge

Shallow discharge is using only small amounts of electricity from the battery before recharging. When we start a car, we are actually connecting the battery. After the car is started, the battery will automatically be recharged as we drive. The shallow discharge is this small use of the battery. Note that doing a shallow discharge necessarily means that we must replace that electricity by recharging the battery as soon as possible. Batteries for solar cells, just like batteries for cars, are designed to be able to do many shallow discharges in a day, assuming the battery is recharged before the next shallow discharge.

Deep Discharge

A deep discharge is a large use of electricity from the battery before being recharged. This can be a large amount used at one time or small amounts used over many days. Either way the concept of a deep discharge is when we take enough energy from the battery before recharging such that the battery is essentially depleted.

Some people classify deep discharge as draining the battery completely. Other people classify deep discharge as draining not completely, but at least up to 80% of capacity. Either way, each deep discharge will permanently damage your battery.

For example, suppose you use a lawn mower which is battery powered. The gauge tells you when you are getting low on power (through a series of colors: green, then yellow, then red). If you operate the mower all the way until it completely dies, then you have created a deep discharge.

Yes you can recharge your battery again, but the recharge is not as effective. Furthermore, if your standard practice is to keep driving the mower until the battery is dead (this would be repeated actions of deep discharge) then you will continue to damage the battery.

Consequently, each recharge after the deep discharge will become less effective. The battery will not be able to recreate the anode as fully, and therefore will only build up a small amount of power no matter how long you charge it.

Manufacturers of batteries will often specify how often you can do this before the battery is no longer useful.

On the other hand, if you stopped mowing at the yellow, and then recharged the battery, the battery would remain undamaged. You could recharge the battery to its fullest level, and start mowing again tomorrow. (See "daily depth of discharge" below).

Note that if you wish to use your battery for long periods before recharging, then this is a deliberate use of deep-charging. If deep discharge is desired then you must buy a battery specifically designed for this purpose. These deep discharge batteries are usually larger, heavier, and more expensive than other batteries.

Self-Discharge

Self-discharge is a slow drain from the battery even though the battery is not actually being used. For example, if we let a car sit for a long time without being used, then the power would slowly drain from the battery until there is no more power in the battery. This is called "self-discharge." Note that all batteries will slowly lose power even if not being used.

Daily Depth of Discharge

Daily depth of discharge is essentially the amount of electricity that we can take from the battery on a daily basis, assuming we recharge the battery soon after, without damaging the battery. This is the same as the daily operation of starting your car. The amount of electricity used in starting your car would be within the manufacturer's limits of daily depth of discharge.

Similarly, using the lawn mower example above, if we stopped using the mower at a specific point in the yellow range (the allowed daily depth of discharge), and took the battery inside to be recharged, then we can operate the lawn mower again tomorrow. We can do this every day (for about 3-5 years) without damaging the battery. Therefore, we are using our battery within the manufacturer's limits of daily depth of discharge, and extending the life of our battery as much as possible.

Maximum Allowable Depth of Discharge

The maximum allowable depth of discharge is the maximum power you can take from the battery at one time without damaging the battery. Suppose that one day you need a greater amount of power from the battery than normal. The maximum allowable depth of discharge is the maximum power you can take at one time, and not reach deep discharge amounts.

The difference between the "daily depth of discharge" and the "maximum allowable depth of discharge" is subtle. Think of it in terms of running. The daily depth of discharge is akin to how far you can run each day without being completely exhausted. You can run the same distance tomorrow. In contrast, the maximum allowable depth of discharge is akin to a mini-marathon. You will be more drained than normal but you will be okay. Just don't try running that same distance tomorrow.

Summary

1. Batteries store energy for later use. We can buy batteries fully charged, or we can use a source such as solar power to charge our batteries.

2. Batteries are essential if the application is to be completely electrically independent. Stand-alone applications are best for smaller power needs such as lighting and communication devices.

3. There are three things needed for a battery to work:
    a. Two metals, each of different materials
    b. Ions in solution or in a paste to create free electrons
    c. A metal wire to conduct electricity for practical use

4. The basic process of the battery is as follows:
    a. The electrolyte solution reacts with the anode. This reaction creates free electrons.
    b. The nuclei of the cathode has a very strong pull which will pull any free electrons which were created at the first metal.
    c. A wire is connected between the two metals which allows the free electrons to travel more easily.
    d. An electrical current is therefore created: the strong nuclei at the cathode will pull on the free electrons at the anode. These free electrons are pulled through the wire, through any device, and ultimately ending at the cathode. This is the electrical current which is created by the battery.

5. Because the two metals are not the same, a voltage difference automatically exists. The difference is primarily due to the pull strength of each metal, and is also related to the kinetic energy of the electrons at each metal. This difference is the voltage stated for the particular battery.

6. The two metals are called the anode and the cathode. The anode is the metal which creates free electrons, and which loses electrons in the circuit. The cathode is the metal which gains electrons in the circuit. Electrons always flow through the wire from the anode to the cathode.

7. The anode is the metal which is progressively being worn away. This is a type of corrosion and has practical significance in pipelines and transmission lines.

8. When we speak of the "size" of a battery, there are really several concepts. A "larger" battery can come in several forms:
    a. A single battery with greater storage capacity
    b. Multiple batteries connected to create a large total capacity
    c. A single battery with a greater depth of discharge

9. The size of the batteries you need will depend on a) how much energy you want to store for later, and b) how independent you wish to be from the utility company.

10. The term "discharge" refers to electricity leaving the battery. There are different types of discharge, depending on the amount of discharge and how often the discharge occurs.

11. The most common types of discharge are:
    a. Shallow discharge: using only small amounts of electricity from the battery before recharging.

    b. Deep discharge: using a large amount of electricity from the battery before recharging. After a deep discharge the battery is essentially depleted.

    c. Self-discharge: slow drain from the battery even though the battery is not actually being used.

    d. Daily depth of discharge: the amount of electricity that we can take from the battery on a daily basis, assuming that we recharge the battery soon after.

    e. Maximum allowable depth of discharge: the maximum power you can take from the battery without reaching deep discharge amounts.

1.7
Resistance and Temperature

Introduction

Resistance and temperature are presented together because these concepts are interrelated and they have similar practical effects. Each factor contributes to power loss. Each factor can damage equipment. Each factor affects the overall efficiency.

List of topics for this chapter
1. Resistance and Temperature Basics
2. Resistance and Temperature on the Atomic Scale
3. Resistance and Heat on the Atomic Scale
4. Factors Affecting Resistance and Temperature
5. Power vs. Power Loss
6. Resistance, Heat, Power Loss, and Temperature

Resistance and Temperature Basics

Resistance is the amount of opposition to the flow of current. Every material has its own amount of electrical resistance. The better conductors have less resistance, but any conductor naturally has some resistance. Wire size is also important in relation to resistance. Generally, larger diameter wires have less resistance while smaller diameter wires have more resistance.

Temperature is a measure of kinetic energy. When molecules move faster they exist at a higher temperature. When molecules move slower they exist at a lower temperature. Temperature is of practical importance because electrical equipment can become seriously damaged when the internal temperature is too high. On a microscopic level, the high temperatures can cause molecular bonds to break. On a visible level, the high temperatures can cause cracks, melting, and electrical fires.

Resistance and Temperature on the Atomic Scale

Temperature is directly related to resistance. In net results, any resistance to electrical current will ultimately increase the overall temperature of the equipment. We can understand the relationships if we examine a wire at the atomic level.

When electrons are moving (either free electrons or the electrons bumping like dominoes), some of those electrons will be impeded by objects. These objects are usually atoms, molecules, and various other electrons. The electrons of the current will hit the object, then take a detour. At this point, the electron will have less kinetic energy, while the object in the way will have more kinetic energy. (This is because the electron has transferred its energy to the impeding object).

Therefore, any object which "resists" the flow of an electron is actually absorbing the kinetic energy of the electron. Then, as discussed earlier, this ultimately means that the object exists at a higher temperature. Therefore, any atom or molecule which resists electrical current will naturally exist at a higher temperature.

Furthermore, the kinetic energy (temperature) of this object will pass on its kinetic energy to neighboring objects (such as other atoms).

Meanwhile, electrical current is still being sent down the wire, and this current continues to meet resistance in that same section of wire. That section of wire continues to get increased kinetic energy...which is again passed onto neighboring objects. Thus, as current continues to travel and meet resistance in a section of wire, the temperature of that section of wire will gradually increase higher and higher.

In this way: 1) any electrical current will be resisted by objects (usually atoms and molecules), 2) this resistance will increase the kinetic energy of the objects (while decreasing the kinetic energy of the electrons), and therefore 3) increase the overall temperature of the wire.

Resistance and Heat on the Atomic Scale

Heat is similar to temperature, but slightly different. Temperature is a measurement of kinetic energy. Heat is the transfer of that energy.

Whenever an atom or molecule gains energy, the atom will want to pass that extra energy onto other atoms and molecules. There are several mechanisms to transfer this energy, but one of the most common ways is through bumping. Electrons, atoms, and molecules with higher kinetic energy will bump into electrons, atoms, and molecules of lower kinetic energy, thereby transferring kinetic energies from one particle to another. This process of energy transfer is what we commonly know as "heat".

Note that the atoms in the wire will transfer their energy to any other atoms or molecules in proximity. For power lines, this usually means passing the energy to air molecules. Therefore, as power lines gain kinetic energy (due to the atoms resisting electrical current), this energy is transferred to the air molecules surrounding the wire. This natural mechanism allows the power lines to keep cool enough to prevent permanent damage. (This process will be discussed in greater detail in the volume on Transmission of Electrical Power).

Therefore, as electrical current passes through a wire, the current will naturally meet some resistance. This resistance will increase the temperature of the wire, which will then create heat as the wire tries to get rid of the excess kinetic energy.

Factors Affecting Resistance and Temperature of Wires

Introduction

The amount of temperature and heat created through a wire depends on the amount of resistance. The amount of resistance is related to three factors: 1) the atomic structure of the material, 2) the amount of current, and 3) the amount of voltage.

Material and Resistance

The amount of resistance is primarily determined by the material. Every material (and hence every wire) has an inherent amount of resistance. This is entirely due to the molecular structure, and how difficult it is for electrical current to pass through the atoms and molecules of the material.

After the wire has been selected and installed, the temperature and heat due to resistance will be determined by the amount of current and the amount of voltage.

Factors Affecting Temperature Due to Current

With a greater current the wire will become a higher temperature. This is because more electrons are bumping per second, and therefore more electrons are meeting resistance per second. Consequently, more atoms and molecules are gaining kinetic energy per second.

Factors Affecting Temperature Due to Voltage

A higher voltage will increase temperature of the impeding molecules. This is primarily due to the energy of impact.

Remember that resistance is the two part action of: 1) an electron hitting an atom or molecule, and 2) transferring kinetic energy to that atom or molecule. Also remember that voltage is the kinetic energy of electrons.

Therefore, when we have a higher voltage, the electrons have higher vibrational energy. Consequently, the impact of a higher energy electron with the impeding object will be much more powerful. (Think of a car crash at higher speeds). This results in the impeding object having much higher kinetic energy, and therefore will exist at a higher temperature.

Temperature of Electrons

The increase in temperature is also due to the inherent nature of voltage. Remember that voltage is the vibrational energy of the electron, and is therefore in fact a type of kinetic energy. Also remember that kinetic energy is a measurement of temperature. Therefore, the amount of vibrational energy of the electrons in the wire could be classified as either "voltage" or "temperature", and measured in either "Volts" or "degrees".

Note that of the two types of kinetic energy, the kinetic energies of atoms tend to produce a greater temperature overall than the kinetic energies of electrons. Yet both can contribute, particularly with very high voltage power lines.

Summary of Factors

Therefore, the amount of total internal temperature of a wire depends mainly on the amount of resistance inherent in the wire. The amount of resistance then depends primarily on the atomic structure of the material, and secondly on either the amount of current or the amount of voltage coming through that wire.

Power vs. Power Loss

Overview of Power and Power Loss

You will often hear the terms "power" in reference to an appliance, and "power loss" in reference to an electrical line. In a scientific sense both terms are describing the same concept. In both cases the same thing happens. However, the key difference is whether we want this phenomenon or not. Here are some key points which clarify the concepts of power, power loss, and resistance:

1. Many electrical devices can be thought of as large resistors. Electrical current often hits particular electrons or molecules in the wires in order to perform the desired function. As described above, this results in resistance. However, that power is used in ways we *do* want, and therefore this resistance, and this power, is the actual use of electrical power we desire.

2. Electrical lines which deliver power to our homes also have resistance. Due to the resistance on the lines, electrical power is used. However, this power is used in ways that we *don't* want, and therefore it is a power loss.

Power Relationships Compared

Earlier we stated that Power = Voltage x Current, and we said that this is the most fundamental relationship in electricity. This is still true, but now that we have discussed resistance we can have a more advanced concept. Electrical power can be related to any two of the following factors: voltage, current, or resistance.

- Power = Voltage x Current   (P = V x I)
- Power (or Power Loss) = Current² x Resistance   (P = I² x R)
- Power (or Power Loss) = Voltage² / Resistance   (P = V² / R)

Whether you use the combinations of a) Voltage and Current, b) Resistance and Current, or c) Voltage and Resistance will depend on the particular situation. Below are some examples.

Power Loss, Resistance, and Current
    Concept: Power Loss = Current² x Resistance
    Symbols: $P = I^2 \times R$
    Units: Watts (of power loss) = (Amps)² x Ohms

    Example: If the current is 7 Amps, and the resistance is 17.14 Ohms, then Power Loss = 7² x 17.14 = 840 Watts.

Power Loss, Resistance, and Voltage
    Concept: Power Loss = Voltage² / Resistance
    Symbols: $P = V^2 / R$
    Units: Watts (of power loss) = (Voltage)² / Ohms

    Example: If the voltage is 120 Volts and the resistance is 17.14 Ohms, then the Power Loss = 120² / 17.14 = 840 Watts.

Resistance, Temperature, Heat, and Power Loss

Tying these concepts together, we can say the following things happen in a wire:

Resistance, Temperature, and Heat

Every wire has an inherent amount of resistance due to the atomic structure of the material. Any object (such an atom or molecule) which resists the flow of an electron is actually absorbing the kinetic energy of the electron. Recall that a higher temperature means that atoms and molecules vibrate faster. Therefore, a collision between an electron of the current and an impeding atom will ultimately result in the impeding atom existing at a higher temperature.

Eventually, these higher energy atoms will want to get rid of their excess energy. Some of this energy will be transferred through heat, where higher energy atoms pass their energy onto lower energy atoms.

If the power line is surrounded by air, then the air molecules will absorb this energy. However, if the power line is contained (such as power lines placed underground) then the heat energy can only transfer to other atoms within the wire...which increases the temperature of the line without any means of cooling. Furthermore, if the atoms of the equipment vibrate with enough energy this may result in cracks, melting, and fires.

Voltage, Current and Temperature

Voltage and current will also increase temperature. Higher voltage electrons will result in higher energy impacts, and therefore the atoms being hit will exist at a higher temperature.

Regarding current, a higher current will result in more collisions per second, and thus more atoms will exist with higher kinetic energy at any moment in time. The net result is an increase in the overall temperature of the atoms in the region.

Resistance and Power Loss

In addition to the temperature increase, resistance to electrical current causes energy loss and power loss. Energy loss occurs because the kinetic energies of the electrons in the electrical current are absorbed by the atoms. These electrons can still bump or flow as current, but they have less energy after impact. Power loss then occurs because power is energy produced or used per time, therefore any energy loss will also result in power loss.

Total Effects

Therefore, any resistance in a power line will result in increases in the amount of all of the following: temperature, heat, power loss, and potential destruction.

Also remember that all power lines and equipment have specified limits on how much current and how much voltage can be put through a wire without damaging the equipment. We must adhere to these limits in order to prevent permanent damage.

Chapter Summary

1. Resistance and temperature each affect how much power we can transmit over wires. Each factor affects the overall efficiency.

2. Resistance is the amount of opposition to the flow of current.

3. When electrical current travels through a wire, that current meets resistance. It is this resistance that creates power loss.

4. Resistance, Current, and Power Loss
    In concept: Power Loss = Current² x Resistance
    In symbols: $P = I^2 \times R$
    In Units: Watts (of power loss) = (Amps)² x Ohms

5. Resistance, Voltage, and Power Loss
    In concept: Power Loss = Voltage² / Resistance
    In symbols: $P = V^2 / R$
    In Units: Watts (of power loss) = (Voltage)² / Ohms

6. We want to make a distinction between the resistance in the appliances and resistance in the electrical lines. This translates into a distinction between desired power (useful power) and undesired power (power loss).

7. Temperature is a measure of the kinetic energy of molecules.

8. All power lines and equipment have temperature limits. Exceeding the temperature limits can damage or destroy the equipment.

9. All resistance to electrical current results in higher temperature atoms and molecules. This is due to the transfer of energy by the electrons hitting atoms in the wire.

1.8
Electrical Connections (Series and Parallel)

Introduction

Every electrical item, from appliances to power plants, can be connected in series, in parallel, or a combination series and parallel. The method of connection can make a practical difference in the amount of current, the amount of voltage, and the amount of power. The method of connection can also make a difference regarding power outages.

List of topics for this chapter
1. Appliances Connected in Series and Parallel
2. Power Sources Connected in Series and Parallel

Appliances Connected in Series and Parallel

Series (Figure: 8.1)

Devices connected in series are connected together directly. Wires out of one device lead into another.

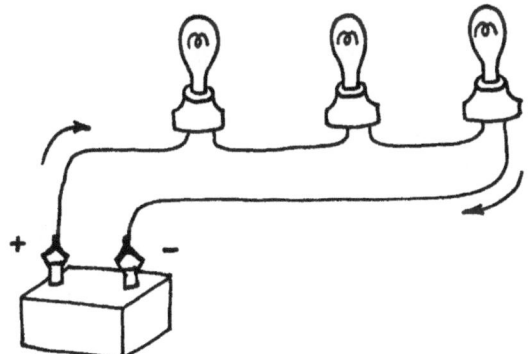

Figure 8.1   Devices Connected in Series

The same current flows through all the devices. However, if one of the devices goes out, then all of the devices beyond that device will go out. Note that if we have a *circuit* connected in series, and one device goes out, then *all* the devices anywhere on that circuit will go out. This is because the circuit was broken and therefore no current can flow anywhere.

For appliances connected in series:

Factor	Through Each Device	Simple Math
Current:	Same in each device	Current in = $I_1 = I_2 = I_3$
Resistance:	Depends on device, Adds for total	$R_{Total} = R_1 + R_2 + R_3$
Voltage:	Divided among each device	$V_{in} = V_1 + V_2 + V_3$

If one device goes out, then all the devices go out.

Parallel (Figure: 8.2)

Devices connected in parallel are more independent. Wires *branch off* from the main line in order to reach each device.

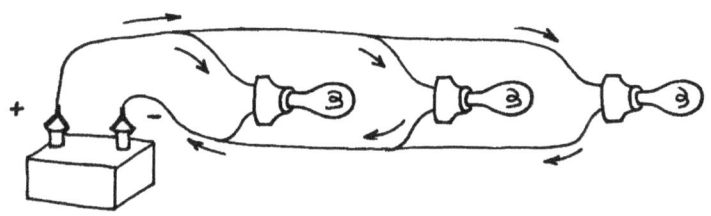

Figure 8.2   Devices Connected in Parallel

Each device is independent of the others. Each device draws its own amount of current. In this situation, if one device goes out then the other devices will still be operational. For appliances connected in parallel:

Factor	Through Each Device	Simple Math
Current:	Divided among each device	Current in = $I_1 + I_2 + I_3$
Resistance:	Depends on device, Adds for total	$R_{Total} = R_1 + R_2 + R_3$
Voltage:	Same in each device	$V_{in} = V_1 = V_2 = V_3$

If one device goes out, that one branch will go out, but the other branches can still supply power.

Series and Parallel together (Figure 8.3)

Devices can also be connected as a combination of series and parallel connections (Figure 8.3). The branches are connected to each other in parallel. Each set of devices on a branch is connected in series. If one device goes out, such as device 1b, then the rest of the devices on that branch will also go out. This is because all the devices on that branch are connected in series.

However, all the other devices on the other branches will still work. This is because the other branches are connected to the power source in parallel.

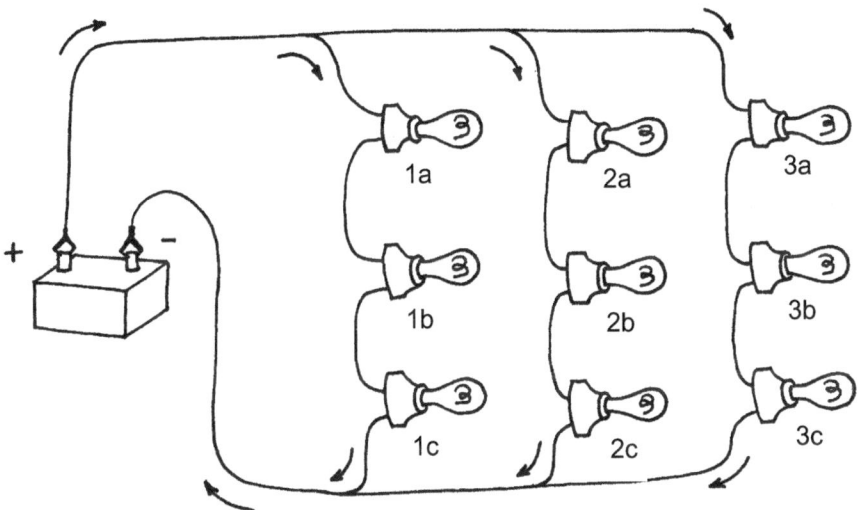

Figure 8.3   Devices Connected in Series and in Parallel

The rules for voltage and current still apply. However there are more factors to consider and so calculating voltage and current at any piece becomes slightly more complex.

Parallel portion
- Voltage in $= V_{branch\ 1} = V_{branch\ 2} = V_{branch\ 3}$
- Current in $= I_{branch\ 1} + I_{branch\ 2} + I_{branch\ 3}$

Series portions
Branch 1 (series connections):
- Voltage$_{branch\ 1,\ in} = V_{1a} + V_{1b} + V_{1c}$
- Current$_{branch\ 1,\ in} = I_{1a} = I_{1b} = I_{1c}$

Branch 2 (series connections):
- Voltage$_{branch\ 2,\ in} = V_{2a} + V_{2b} + V_{2c}$
- Current$_{branch\ 2,\ in} = I_{2a} = I_{2b} = I_{2c}$

Branch 3 (series connections):
- Voltage$_{branch\ 3,\ in} = V_{3a} + V_{3b} + V_{3c}$
- Current$_{branch\ 3,\ in} = I_{3a} = I_{3b} = I_{3c}$

Power Sources Connected in Series and Parallel

Introduction

In this section we will look at connecting multiple power sources. Connected power sources may include batteries, solar cells, wind turbines, and large power plants. The basic principles of series connections and parallel connections still apply. However, because we are looking at the sources of power rather than the uses of power the situation is slightly different. In this section we look at the same principles of series and parallel connections, but specifically applied to connecting multiple power sources.

Series Connection for Power Sources (Figure: 8.4)

Power sources connected in series are connected together directly: wires out of one power source lead into another. The basic principles of series connections still apply. When power sources are connected in series the voltage adds, while the current is the same through each source.

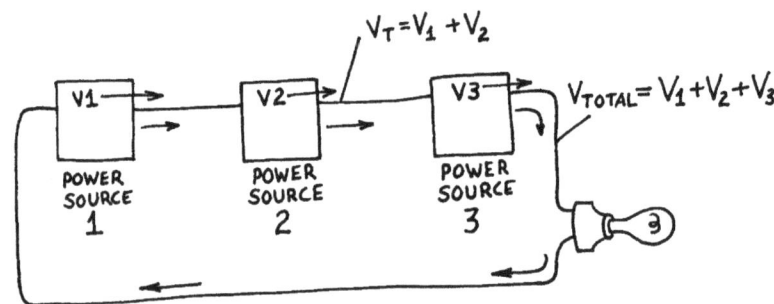

Figure 8.4   Power Sources Connected in Series

The primary advantage to connecting power sources in series is the increase in voltage. When power sources are connected in series the voltage adds. The total voltage is the sum of all the voltages of each power source.

The primary disadvantage to connecting power sources in series is significant loss of power due to any failure in the system. In a series connection the same current flows through all the power sources. Therefore, if one of the power sources goes out then less power will be sent. For example, in the diagram above if power source #3 goes out then the power from sources from #1 and #2 will not be sent either.

The series connections for power sources can also lead to cascading power outages. For example, if power source #3 goes out, this can cause a reverse of power flow to power source #2. This collision of electrical power can create a voltage spike which will damage power source #2. Now we have two power sources out.

In addition, the power from failed sources #3 and #2 can reverse flow to #1, thereby causing another voltage spike and another failed power source. In this way, when power sources are connected in series, a failure in one device can cause cascading power outages throughout the system. (See the volume on Grids in this series for more details).

For Power Sources connected in Series:

Factor	From Each Power Source	Simple Math
Current:	Same from each source	Current Out = $I_1 = I_2 = I_3$
Voltage:	Adds for total	Voltage Total = $V_1 + V_2 + V_3$

If one power source goes out, then any power from previous power sources will not be sent. In addition, reverse power flow, voltage spikes, and cascading power outages are also possible.

Parallel connections for power sources (Figure: 8.5)

Power sources connected in parallel are more independent of each other. Wires *branch off* from the main line in order to reach the final device. The basic principles of parallel connections still apply. When power sources are connected in parallel the currents add but the voltages do not add.

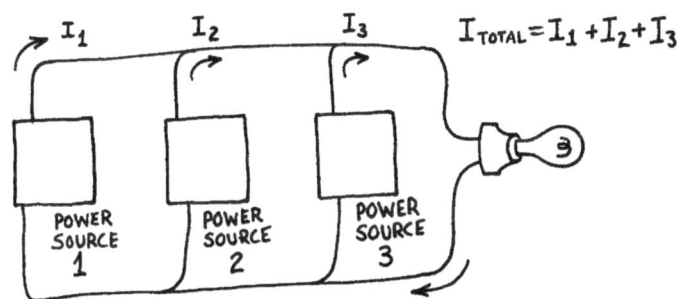

Figure 8.5   Power Sources Connected in Parallel

When power sources are connected in parallel, each source is independent of the others. Each source has its own amount of current. In this situation, if one power source goes out, then it is more likely that the other power sources will still be operational.

Factor	From Each Power Source	Simple Math
Current	Adds for total	Current Total = $I_1 + I_2 + I_3$
Voltage	Same from each source	Voltage Out = $V_1 = V_2 = V_3$

Summary

1. Every electrical item can be connected in series, connected in parallel, or connected as a combination of series and parallel.

2. Devices connected in Series:
    - Current: same through each device
    - Voltage: divided among each device
    - Power outage: When one device goes out, all the devices will go out

3. Devices connected in Parallel:
    - Current: divided among each device
    - Voltage: same in each device
    - Power outage: When one device goes out, the others will still work

4. Power Sources connected in Series:
    - Current: same in each device
    - Voltage: adds for total
    - Power Outage: When one power source goes out, less power sent, also possible cascading power failures
    - Advantage: increased voltage

5. Power Sources connected in Parallel:
    - Current: adds for total
    - Voltage: same in each device
    - Power Outage: When one power source goes out, the other power sources are likely to still work
    - Advantage: power failures are less likely

1.9
Other Primary Electrical Terms

Introduction

There are many electrical terms that you may come across when reading books and reports on electricity. This chapter offers the most important of all the other electrical terms that you may come across in your future readings.

List of topics for this chapter
1. Conductor, Resistor, Resistance, Insulator
    - Conductor
    - Resistor and Resistance
    - Insulator
    - Resistor vs. Insulator
    - Impedance
    - Reactance

2. Inductor, Inductance, Capacitor, Capacitance
    - Inductor and Inductance
    - Capacitor and Capacitance
    - Inductor and Capacitor – Special Relationships

3. Rotor, Stator, Electric Motor
    - Rotor and Stator
    - Electric Motor

4. Conversion Equipment
    - Diode
    - Power Supply

Conductor, Resistor, Insulator

Conductor

A conductor is any wire which allows the flow of electricity. Metals make the best electrical conductors. Commonly used materials for conductors include: copper, aluminum, iron (or steel), and gold. Water and many soils are reasonable conductors, due to the naturally occurring ions. Note that air is not a good conductor.

Resistor and Resistance

A resistor is any electrical component which interferes with the flow of electrical current. Resistance is the amount of opposition to that current.

Every material has its own amount of electrical resistance. The better conductors have less resistance, but any conductor naturally has some resistance. There are actually three types of resistance (Resistance, Reactance and Impedance). We will discuss these types of resistance later in the sections on reactance and impedance.

Insulator (Electrical Insulator)

An insulator is any material which prohibits electricity from flowing. We put insulators around our wires to prevent electricity traveling where we don't want it to go. Good electrical insulators include: air, ceramics, oxides, plastics, and rubber.

Special note: Resistor vs. Insulator

A resistor is like a speed bump. We *want* the electricity to travel in that direction, but the resistance gets in the way. The electricity can still travel that direction, but not with as much energy as before. The insulator is more like a brick wall. The electricity is physically prohibited from going that direction and must turn around.

Impedance

Impedance, at its most basic, is a type of resistance. Specifically, impedance is the combination of three types of resistance:
1. Resistance through a Resistor ("Resistance")
2. Resistance through an Inductor ("Inductive Reactance")
3. Resistance through a Capacitor ("Capacitive Reactance")

See the next chapter on Impedance and Reactance for more details.

Reactance

Reactance is a type of resistance. Reactance is the amount of opposition to current as the current goes through an Inductor, through a Capacitor, or through both. Note that the term Reactance is specific to situations where the current goes through an inductor or a capacitor, not a conductor.

Inductor/Inductance, Capacitor/Capacitance

Inductor and Inductance

Definition pair #1 for Inductor and Inductance:
- An Inductor is any device which creates electrical current
- Inductance is the creation of electrical current

Definition pair #2 for Inductor and Inductance:
- An Inductor is also any device which pulls current forward
- Inductance is the amount of current being pulled forward

We have talked about inductors on a large scale – these are the devices we call generators. However, many electronic devices such as radios and video games will have small inductors built into the circuitry. Inductors play many important roles inside these electronic devices. Furthermore, inductance can occur in many points along the path of electrical power. This inductance can produce undesired effects. (See the section on "Inductor and Capacitor Special Relations" below.)

Capacitor and Capacitance

Definition pair #1 for Capacitor and Capacitance:
- A Capacitor is any device which holds electrical charge
- Capacitance is the amount of charge held by the capacitor

Definition pair #2 for Capacitor and Capacitance:
- A Capacitor is any device which holds current back
- Capacitance is the amount of current is that is held back

Inductor and Capacitor – Special Relations

When you have both an inductor and a capacitor in one circuit there is a tug of war between them. The inductor pulls current forward while the capacitor holds current back. Changing the amount of inductance relative to the amount of capacitance can create many practical results. The specific result will depend on the relative strengths of the inductor and the capacitor, and how they are connected. We then use these results to our advantage.

There are two main areas where the relationship between inductor and capacitor must be considered: power lines and circuitry. In power lines, the inductor-capacitor relationship has a significant effect on efficiency. Electrical current can act like a horse: sometimes it goes faster than we want, sometimes it goes slower. We need to rein the current in or get it moving faster as the situation requires. The cause is usually too much inductance in the line, which then makes current go faster than we want, and so we rein it in using a capacitor. (For more details see the later sections in this series on capacitor switching and on power factor).

In circuitry, the inductor-capacitor relationship is the basis for many of the practical results seen in devices. However, we will not discuss such circuitry in this book.

Rotor, Stator, Electric Motor

Rotor and Stator

The rotor and the stator are two key parts of an electrical generator. The *rotor* is the part of the generator that *rotates*. In our generator examples, the rotor is the magnet. The *stator* is the part of the generator that is *stationary*. In our generator examples, the stator is the set of wires.

Electric Motor

An electric motor is any device which turns electrical energy into mechanical energy. The term "electric motor" is a broad term. This includes motors which run on batteries (DC) and motors which are plugged into the wall (AC). This term also includes subsets of motors such as induction motor and synchronous motor. The primary types of electric motors are described and clarified below.

induction motor

The induction motor is an electrical motor which specifically runs on alternating current. Induction motors have magnets, just as generators do. Notice that any device which a) has a motor, and b) plugs into the wall, uses an induction motor.

Essentially, the induction motor works by being the reverse of the electrical generator. In the electrical generator mechanical energy is converted to alternating current through the use of a magnet. In the induction motor the alternating current is converted back into mechanical energy also by using a magnet.

The mechanical energy from the induction motor is usually produced as rotational motion. This rotational motion then operates gears, axels, etc. in order to perform specific mechanical functions.

synchronous motor

The synchronous motor is a type of induction motor. The term synchronous motor specifically refers to an induction motor in which the speed of the motor is proportional to the frequency of the alternating current coming into the motor.

Conversion Equipment

Diode

A diode is an electronic one-way street. A diode allows current to flow in only one direction.

Power Supply

A power supply modifies the electrical power coming from the outlet into a suitable form for whatever equipment we are using. The power supply does one or more of the following: changes the voltage, converts AC to DC, and/or changes the frequency.

The power supply that most of us are familiar with is the equipment used to recharge batteries. Many electronic devices today use rechargeable batteries, including devices such as cell phones and electric razors. When we plug such a device into the wall to recharge the batteries, we are in fact using a power supply.

For example, let us recharge a cell phone. Electricity coming to the outlet is AC, but a battery is DC. Therefore, the power supply must convert AC to DC in order to recharge the battery. The power supply for our cell phone also does something else: it reduces the voltage. The electricity coming to the outlet is 120 Volts, but the cell phone uses only about 5 volts. Therefore, the power supply reduces the voltage to a level suitable for the cell phone.

Do not confuse a "power supply" with a "power source." The power supply does not create power, rather the "power supply" modifies the power for a specific use. If you want to refer to the generation of power (such as coal power) then the proper term to use is "power source."

Chapter Summary

1. <u>Conductor</u> – any wire which allows the flow of electricity

   The best electrical conductors include: copper, aluminum, iron, steel, and gold. Water and many soils conduct electricity reasonably well due to the naturally occurring ions.

2. <u>Resistor</u> – any electrical component which interferes with the flow of electricity. Any conductor is also a resistor to some degree.

3. <u>Resistance</u> – the amount of opposition to the flow of current.

4. <u>Insulator</u> – any material which prohibits electricity from flowing. Good electrical insulators include: air, ceramics, oxides, plastics, and rubber.

5. <u>Inductor</u> – any device which creates electrical current, or any device which pulls current forward.

6. <u>Inductance</u> – the creation of electrical current, or the amount that a current is pulled forward.

7. <u>Capacitor</u> – any device which holds electrical charge, or any device which holds current back.

8. <u>Capacitance</u> – the amount of charge held by the capacitor, or the amount that a current is held back.

9. <u>Inductor and Capacitor – Special Relations</u>

   When you have both an inductor and a capacitor in one circuit, there is a tug of war between them. The inductor pulls current forward, while the capacitor holds current back.

10. <u>Impedance</u> – a type of resistance; specifically, the combination of three types of resistance:
    a. Resistance through a Resistor ("Resistance")
    b. Resistance through an Inductor ("Inductive Reactance")
    c. Resistance through a Capacitor ("Capacitive Reactance")

11. <u>Reactance</u> – Reactance is a type of resistance, specifically used to refer to where current goes through an Inductor or through a Capacitor.

12. <u>Rotor and Stator</u> – two key parts of an electrical generator: The rotor is the part of the generator that rotates. The stator is the part of the generator that is stationary.

13. <u>Diode</u> – any electronic component which allows current to flow in only one direction.

14. <u>Power Supply</u> – any device which modifies the electrical power from the outlet into a suitable form for whatever equipment we are using. The power supply does one or more of the following: changes the voltage, converts AC to DC, and/or changes the frequency.

15. <u>Electrical Motor</u> – any device which turns electrical energy into mechanical energy. This term can refer to motors run by direct current or run by alternating current.

16. <u>Induction Motor</u> – an electrical motor which is powered by alternating current. Any device which has a motor and plugs into the wall is a device which uses an induction motor. The induction motor converts alternating current into mechanical energy by using a magnet. The action of the induction motor is essentially the reverse of an electric generator.

17. <u>Synchronous Motor</u> – an induction motor in which the speed of the motor is proportional to the frequency of the incoming alternating current.

1.10
Impedance and Reactance Details

Introduction

Impedance and Reactance are of practical importance to electrical power because these factors contribute to power loss, and adjusting these factors can achieve greater efficiency.

However, these factors require more explanation than the previous electrical terms, and the calculations require more explanation, therefore these concepts require their own separate chapter.

List of topics for this chapter
1. Impedance: Basic Concepts
2. Resistance, Reactance, and Impedance Specifics
3. Symbols: L, C, R, X, and Z
4. Calculations of Reactance and Impedance

Impedance: Basic Concepts

Introduction

Impedance, at its most basic, is a type of resistance. Impedance is in fact the *total* resistance. In our first level of understanding we said the following two things: 1) a resistor is any electrical component which interferes with the flow of electrical current, and 2) resistance is the amount of opposition to the flow of electrical current.

Now we go deeper in our understanding. There are three objects that cause resistance: Resistor, Inductor, and Capacitor. Thus, there are three types of resistance, each type of resistance being based on the particular object which caused the interference:

1. Resistance through a <u>Resistor</u> or Conductor ("Resistance")
2. Resistance through an <u>Inductor</u> ("Inductive Reactance")
3. Resistance through a <u>Capacitor</u> ("Capacitive Reactance")

The terms resistance, inductive reactance, and capacitive reactance, as well as the term impedance, are each similar in the following ways:
   a) each is a measurement of the opposition to electric current
   b) each term is measured in Ohms.

However, the terms have subtle differences: The term "reactance" is used specifically for resistance through an inductor or resistance through a capacitor. The term "resistance", when used in the same report as reactance, refers only to the resistance in the conductor (aka resistor). Finally, "impedance" is the total of all the resistances.

Resistance, Reactance, and Impedance Specifics

1. <u>Resistance</u>: The amount of opposition to current flow that goes through a resistor. Symbol: R. Note that all conductors will have some natural resistance. Therefore, resistance through a conductor is included at this point.

2. <u>Reactance</u>: The amount of opposition to current flow that goes through inductors, capacitors, or both.
   a. <u>Reactance by Inductor</u> (inductive reactance): The tendency of an Inductor to resist current flowing through it. Symbol X_L
   b. <u>Reactance by Capacitor</u> (capacitive reactance): The tendency of a Capacitor to resist current flowing through it. Symbol X_C
   c. <u>Total Reactance</u>: The total resistance to current flow through both the inductor and the capacitor. Symbol X.

3. <u>Impedance</u>: Impedance is the total opposition to current flowing through the circuit. Impedance is the combination of: resistance, reactance by the inductor, and reactance by the capacitor. Symbol: Z
   Note that many technicians and books shorten the definition of impedance by saying: Impedance is the combination of Resistance and Total Reactance. Just remember when you read such a statement that Total Reactance is actually two types: inductive reactance and capacitive reactance.

Symbols: L, C, R, X, and Z

You will often see these symbols thrown about, not just in circuitry diagrams but in general discussions of electrical issues. Therefore the following is a reference list of symbols and their meanings.

L – Inductor, or relating to inductance
C – Capacitor, or relating to capacitance
R – Resistance
X – Total Reactance
X_L – Inductive Reactance (resistance through inductor)
X_C – Capacitive Reactance (resistance through capacitor)
Z – Impedance (total resistance)

Calculations of Reactance and Impedance

Introduction

I do not expect this audience to become experts in circuitry. However, it is worthwhile to know the basics of the calculations for reactance and for impedance because these relationships can greatly influence the efficiency of electrical power.

Calculation of Inductive Reactance

Inductive Reactance, X_L, is related to two things:
1. Size of the inductor (known as inductance)
2. Frequency of the alternating current going through the inductor

The formula for Inductive Reactance is $X_L = 2fL$, where:
X_L is the Inductive Reactance, measured in Ohms
f is the frequency of alternating current, measured in Hertz
L is the inductance, measured in Henrys

Calculation of Capacitive Reactance

Capacitive Reactance is related to two things:
1. Size of the capacitor (known as the capacitance)
2. Frequency of the alternating current going through the capacitor

The formula for Capacitive Reactance is $X_C = (1/2)fC$, where:
X_C is the Capacitive Reactance, measured in Ohms
f is the frequency of alternating current, measured in Hertz
C is the capacitance, measured in Farads

Calculations of Impedance

Combining resistances to find impedance is not as simple as straight addition. This is due the inherent nature of electricity through circuits. However, the calculation is not too difficult.

Impedance is calculated based on using vectors. These vectors form a right triangle. Then, because we have a right triangle we can use the Pythagorean Theorem. The Pythagorean Theorem is $a^2 + b^2 = c^2$. In our situation, "a" is Resistance (R), "b" is the Total Reactance (X), and "c" is the Impedance (Z). This becomes: $Z^2 = R^2 + X^2$. At this point the value for impedance, Z, can be easily solved.

Calculations of Total Reactance

Total Reactance, X, is not a simple addition of inductive reactance and capacitive reactance (although it is related to addition). This is due the nature of inductor and capacitor. Remember that a tug of war exists between the inductor and capacitor. Therefore, combining reactances is conceptually combining two opposites. We are thus adding a negative, where the negative represents the "opposite factor." Mathematically, this is the same as subtraction. Therefore, you subtract one reactance from the other: Total Reactance = $(X_L - X_C)$ or $(X_C - X_L)$.

Note that it does not matter which reactance you subtract from because in the *impedance* calculation (the ultimate goal), the value is squared, and any negative becomes a positive anyway.

The full calculation for Impedance would then look like this:
$$Z^2 = R^2 + (X_L - X_C)^2 \text{ or } Z^2 = R^2 + (X_C - X_L)^2$$

Chapter Summary

1. Reactance and impedance are important to understand because these relationships can greatly influence the efficiency of electrical power.

2. Impedance is the total opposition to current flowing through the circuit. Impedance is the combination of all resistance in a circuit or wire (resistance through a Conductor, an Inductor, and a Capacitor).

3. The terms Resistance, Inductive Reactance, Capacitive Reactance, and Impedance are each a measurement of the opposition to electric current. Each is measured in Ohms.
    • Resistance, R: the amount of opposition to current flow that goes through a Resistor.
    • Reactance, X_L, X_C, X: the amount of opposition to current flow that goes through Inductors, Capacitors, or both.
    • Impedance, Z: the total opposition to current flowing through the circuit. Impedance is the combination of: Resistance, Reactance by the inductor, and Reactance by the capacitor.

4. Full Calculation of Impedance: $Z^2 = R^2 + (X_L - X_C)^2$

Conclusion

Many Americans hold passionate views about electrical power, yet few Americans understand all the details behind their passion. Electricity should not be mysterious. The science, the technology, and the data of electrical power can be understood by anyone.

Above all else, we must remember that there are no perfect solutions, there are only choices. Any option can be beneficial, yet each option has its own technical issues to work with. It is up to you and to your community to make those educated decisions. I hope that this book will help guide you in your choices.

M.F.

Appendices for Introduction to Electrical Power

A1 Energy Unit Equivalents
A2 Power Unit Equivalents
A3 Power, Voltage, Current, & Resistance Equations
A4 Flow Unit Equivalents
A5 Temperature Conversions
A6 Radiation Unit Equivalents
A7 Ash Data
A8 Wire Sizes

A1: Energy Unit Equivalents

A. <u>eV is equivalent to:</u>
1 eV $= 1.6 \times 10^{-19}$ J $= 3.83 \times 10^{-20}$ cal
$= 1.52 \times 10^{-22}$ BTU $= 4.45 \times 10^{-26}$ kW·hr

1 MeV $= 1 \times 10^{6}$ eV $= 1.6 \times 10^{-13}$ Joules $= 3.83 \times 10^{-14}$ cal
$= 1.52 \times 10^{-16}$ BTU $= 4.45 \times 10^{-20}$ kW·hr

B. <u>Joule is equivalent to:</u>
1 Joule $= 6.242 \times 10^{18}$ eV $= 6.242 \times 10^{12}$ MeV $= .2389$ cal
$= 9.48 \times 10^{-3}$ BTU $= 2.778 \times 10^{-7}$ kW·hr

1 kiloJoule $= 1{,}000$ J $= 1$ kW·sec

C. <u>calorie is equivalent to:</u>
1 calorie $= 2.19 \times 10^{19}$ eV $= 4.186$ J $= .0039$ BTU $= 1.163 \times 10^{-6}$ kW·hr
1 kilocal $= 1{,}000$ cal $= 1.63 \times 10^{-3}$ kW·hr

D. <u>BTU is equivalent to:</u>
1 BTU $= 6.58 \times 10^{21}$ eV $= 1{,}055$ J $= 1.055$ kJ $= 252$ cal $= 2.930 \times 10^{-4}$ kW·hr

E. <u>kilowatt-hour is equivalent to:</u>
1 kw-hr $= 2.25 \times 10^{25}$ eV $= 3 \times 10^{6}$ J $= 3{,}600{,}000$ Joules
$= 3{,}600$ kilojoules $= 8.6 \times 10^{5}$ cal $= 3{,}413$ BTU

A2: Power Unit Equivalents

A. <u>Watts or kilowatts are equal to:</u>
1 Watt = 1 Joule /second = 1 kg x m²/s³
= 3.413 BTU/hr = .2389 cal/sec = 0.00134 H.P.

1 kilowatt = 1,000 Watts = 3,413 BTU/hr = 238.9 cal/sec
= 1.34 H.P. = .2389 kilocal/sec

B. <u>BTU/hr are equal to:</u>
1 BTU/hr = .2930 Watts = .07 cal/sec
= .000393 H.P. = .000293 kW

C. <u>cal/sec or kilocal/sec are equal to:</u>
1 cal/sec = 14.22 BTU/hr = 4.186 Watts = .0056 H.P. = .004186 kW
1 kilocal/sec = 5.615 H.P. = 4.186 kilowatts

A3: Power, Voltage, Current, and Resistance

1. <u>Power, Voltage, and Current</u>
   a) in words: Power = Voltage x Current
   b) in symbols: P = V x I
   c) in units: Watts = Volts x Amps

2. <u>Voltage, Current, and Resistance</u>
   a) in words: Voltage = Current x Resistance
   b) in symbols: V = I x R
   c) in units: Volts = Amps x Ohms

3. <u>Power Loss, Current, and Resistance</u>
   a) in words: Power Loss = current squared x resistance
   b) in symbols: P = I² x R
   c) in units: Watts = Amps squared x Ohms

4. <u>Power Loss, related to Voltage and Resistance</u>
   a) in words: Power Loss = Voltage squared / resistance
   b) in symbols: P = V²/R
   c) in units: Watts = Volts squared / Ohms

A4: Flow Unit Equivalents

A. <u>1 gallon/min is equal to</u>:
1 gal/min = 1.0 gal/min = .133 ft^3/min
         = .063 kg/s = .063 L/s
         = .002 ft^3/s = .000062 m^3/s

B. <u>1 ft^3/min is equal to</u>:
1 ft^3/min = 7.48 gal/min = 1.0 ft^3/min
         = .471 kg/s = .471 L/s
         = .016 ft^3/s = .000471 m^3/s

C. <u>1 kg/s is equal to</u>:
1 kg/s   = 15.9 gal/min = 2.12 ft^3/min
         = 1.0 kg/sec = 1.0 L/sec
         = .035 ft^3/s = .001 m^3/s

D. <u>1 L/s is equal to</u>:
1 L/s   = 15.9 gal/min = 2.12 ft^3/min
         = 1.0 kg/sec = 1.0 L/s
         = .035 ft^3/s = .001 m^3/s

E. <u>1 ft^3/s is equal to</u>:
1 ft^3/s   = 448.8 gal/min = 60 ft^3/min
         = 28.5 kg/s = 28.5 L/s
         = 1.0 ft^3/s = .0285 m^3/s

F. <u>1 m^3/s is equal to</u>:
1 m^3/s   = 15,840 gal/min = 2,118 ft^3/min
         = 1,000 kg/s = 1,000 L/s
         = 35.3 ft^3/s = 1.0 m^3/s

A5: Temperature Conversions

<u>Conversion Formulas: Celsius and Fahrenheit</u>
A. Celsius to Fahrenheit: °F = [9/5 x °C] + 32
B. Fahrenheit to Celsius: °C = 5/9 x [°F − 32] or °C = [5/9 x °F] − 17.77

<u>Conversion Formulas: Kelvin, Celsius and Fahrenheit</u>:
A. Celsius to Kelvin: K = °C + 273.16
B. Kelvin to Celsius: °C = K − 273.16
C. Fahrenheit to Kelvin: K = [5/9 x °F] + 255.39
D. Kelvin to Fahrenheit: °F = [9/5 x K] − 459.68

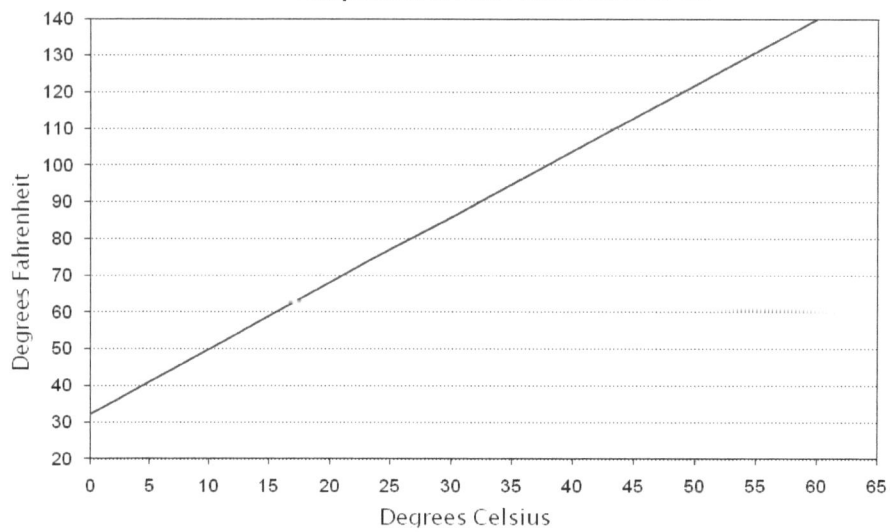

A6: Radiation Unit Equivalents

1 gray = 100 rad = 1 Joule absorbed/kg tissue
1 rad = .01 gray = .01 Joule absorbed/kg tissue

1 Sievert = 1 gray x equivalent dose factors
1 rem = 1 rad x equivalent dose factors

1 Sievert = 100 rem   or   1 rem = .01 Sievert

A7: Ash Data

Molecule	Weight % of molecule in the ash
1. SiO_2 (Silicon Dioxide)	30% – 50%
2. Al_2O_3 (Aluminum Oxide)	10% – 25%
3. Fe_2O_3 (Iron oxide)	5% – 10%
4. CaO (Calcium Oxide)	2% – 25%
5. MgO (Magnesium Oxide)	1% – 6%
6. Na_2O (Sodium Oxide)	.2% – 1.5%
7. K_2O (Potassium Oxide)	1% – 3%
8. TiO (Titanium Oxide)	1%

Note that the exact percentages depend on the region where the coal came from. The percentages above are the averages for coal ash from all regions in the United States.

A8: Wire Sizes

Am. Wire Gauge #		Diameter, (inches)	Area (in^2)	Area (cir. mils)
0000	(4/0)	.4600	.1662	211,600
000	(3/0)	.4096	.1318	167,800
00	(2/0)	.3648	.1045	133,100
0	(1/0)	.3249	.0829	105,600
1		.2893	.0657	83,690
2		.2576	.0521	66,360
3		.2294	.0413	52,620
4		.2943	.0328	41,740
5		.1819	.0259	33,090
6		.1620	.0206	26,240
7		.1443	.0163	20,820
8		.1285	.0123	16,510
9		.1144	.0103	13,090
10		.1019	.0081	10,380
11		.0907	.0065	8,230
12		.0808	.0051	6,530
13		.0720	.0040	5,180
14		.0641	.0032	4,110
15		.0571	.0025	3,260

Bibliography

A. Electrical Principles
(Generators, Turbines, Transmission Lines, Home Wiring)

1. Electrical Power: Motors, Controls, Generators, Transformers, by Joe Kaiser. Publisher: The Goodheart-Willcox Company, Inc.
2. Electricity: Power Generation And Delivery, Sixth Edition, by Alerick and Keljik, 1996. Publisher: Delmar
3. Electricity: Motors, Controls, Alternators, Fifth Edition, by Alerick and Keljik, 1991. Publisher: Delmar
4. The Making of the Electrical Age, by Harold Sharlin, 1963. Abelard-Schuman
5. The Fantastic Inventions of Nikola Tesla, by Nikola Tesla and David Childress, 1993. Publisher: Adventures Unlimited
6. Electrician's Exam Preparation Guide, by John Traister, 2001. Craftsman Book Company
7. The Silent Energy, by Kogan and Pick, part of the "Foundations of Science Library," 1966. Publisher: Greystone Press.
8. Networks of Power: Electrification in Western Society, 1880-1930, by Thomas Hughes, 1983. Publisher: Johns Hopkins University Press.
9. The Lineman's and Cableman's Handbook, Ninth Edition, by Kurtz, Shoemaker, and Mack, 1998. McGraw-Hill
10. Wiring Essentials, by various contributors, Black and Decker Quick Steps Series, 1996. Creative Publishing International.
11. Home Electrical Wiring Made Easy, by Robert Wood, 1993. Publisher: Tab Books, a division of McGraw-Hill
12. Basic Home Wiring, by various contributors, 1989. Publisher: Sunset Books
13. Basic Wiring, by various contributors, 1994. Publisher: Time-Life Books
14. Advanced Wiring, by various contributors, 1998. Publisher: Time-Life Books
15. EMF and Power Lines - Report for the public: *"Electric and Magnetic Fields Associated with the Use of Electric Power"*, EMF RAPID, of National Institute of Environmental Health Sciences, 2002. www.niehs.nih.gov/emfrapid/booklet/

Other Books by Mark Fennell

Paperback Books

Hydropower Technologies Explained Simply
https://www.createspace.com/3956806

Wind Power Technologies Explained Simply
https://www.createspace.com/3957117

Solar Power Technologies Explained Simply
https://www.createspace.com/3985021

Coal Power Technologies Explained Simply
https://www.createspace.com/4007151

Natural Gas and Other Hydrocarbon Technologies Explained Simply
https://www.createspace.com/4009353

Transmission of Electrical Power Explained Simply
https://www.createspace.com/3998474

Utility Operations and Grid Systems Explained Simply
https://www.createspace.com/4004832

e-Books

Practical Considerations of Solar Power
Advanced Solar Cell Technologies
Formation and Mining of Coal
Clean Coal Technologies
Mercury and Coal Power
Health Hazards of Radioactive Decay
Radiation Measurements
Processes of Radioactive Decay and Storage of Nuclear Waste
Extracting and Refining Natural Gas (includes Fracking)
Transportation, Storage, and Use of Natural Gas
Underground Cables
Utility Operations and Quality Control
Power Grids Explained Simply

Index

alternating current: 12, 34-40, 44-45, 49-57, 66-68
amps: 18-19, 45-47, 82, 108
batteries:
    anode and cathode: 62-65
    basic operation of batteries: 60-62
    battery versus turbine-generator: 60, 66-68
    depth of discharge: 71-73
    recharging battery: 65-66, 96
    rust and corrosion: 64-65
    size of batteries: 70-73
BTU: 13-17, 107-108
capacitor: 91-94, 99-102
conductor and conductivity: 43-45, 61-62, 77, 92-93, 99-100
converting energy:
    converting magnetic energy into electrical energy: 31-40
    converting stored energy into flow energy: 12, 19-20, 26-31
current:
    alternating current: 12, 34-40, 44-45, 49-57, 66-68
    direct current: 12, 43-44, 60-63, 66-69, 96
    basic concepts: 18, 43-45, 60-63, 66-69, 108
    generating: 12-13, 18, 20-21, 25, 31-40
    power and current: 11, 18-19, 46, 51, 81-82, 108
    resistance: 77-79, 81-82, 92-93, 99-102, 108
    series and parallel: 85-89
    sine waves: 37-40, 50-51, 53-57
    temperature and current: 79-80
diode: 96
discharge (types of battery discharge): 71-73
electron-volt: 13-14, 17, 47, 107
energy:
    basic concepts: 13-15
    electrical energy: 11, 18, 20, 34-40
    electrode energy (batteries): 63-64, 66-67, 70-73
    kinetic energy: 12, 19-20, 25-31, 77-83, 109
    magnetic energy: 21, 32-40
    mechanical energy: 20, 94-95
    voltage energy: 18, 46-47
    units of energy defined and compared: 13-15, 107

equations:
    current, voltage, resistance, power, and power loss: 18, 81–82, 108
    impedance and reactance: 101–112
    series and parallel: 86–89
    temperature: 110
    units compared: 107–100
frequency: 49–51, 95–96, 101–102
generators:
    alternating power, sine waves, and generators: 35–40, 49–51, 53–58
    basic operation: 11–13, 19–21, 25, 31–40
    batteries versus generators: 66–68
    frequency and generators: 49–51
    magnets and generators: 20–21, 25, 31–40
    motors and generators: 94–95
    turbines and generators: 11–13, 21, 25
IGCC turbine: 31
impedance: 92, 99–102
induction: 34–39, 94–95
inductor: 92–94, 99–102
Joule: 13–15, 46–47, 107
Kilowatt (unit of power): 16–18, 108
Kilowatt-hour (unit of energy): 13–15, 107
kinetic energy:
    kinetic energy of atoms and electrons in wire: 77–83
    kinetic energy of flow of molecules in turbine: 12, 19–20, 25–31, 109
    kinetic energy and temperature: 77–82
magnet and magnetic field: 19–21, 25, 31–40, 49–50, 53, 67–68, 94–95
motor: 94–95
parallel connections: 85–89
power:
    basic concepts of power: 11–12, 16–19, 46, 49–51
    equations: 18, 81–82, 108
    units defined and compared: 16–17, 108
power loss: 77, 81–83, 99, 108
power supply: 96
reactance: 92–93, 99–102
resistor and resistance: 77–82, 86, 92–93, 99–102, 108
rotor and stator: 94

series connections: 85–89
sine wave: 37–40, 50–51, 53–57
temperature: 77–83, 110
three phase electrical supply: 38–40, 49
turbines:
    basic principles: 11–12, 19–21, 25–30
    dual turbine: 30–31
    gas turbine: 29
    IGCC turbine: 31
    steam turbine: 11–13, 28–29
voltage:
    basic concepts: 18, 46–48, 67–68
    batteries: 60, 63–64, 67–68, 70
    equations: 18, 81–82, 107–108
    and power: 11, 18, 46, 51, 81–83, 96
    series and parallel: 85–89
    sine waves: 37–40, 51, 53–56
    and temperature: 80–83
Volts & electron volts: 14, 17–18, 46–48, 107
Volt-ampere: 17
Watts: 16–19, 82, 107

www.ingramcontent.com/pod-product-compliance
Lightning Source LLC
Chambersburg PA
CBHW051019180526
45172CB00002B/403